Return to Roxby Downs

Return to Roxby Downs

John Showers

BOOKENDS

First published 1999

Showers, John
Return to Roxby Downs

ISBN 0 646 38218 7

National Library of Australia
Cataloguing-in-publication entry
1. Showers, John, 1926- . 2. Olympic Dam Project - History.
3. Civil engineers - South Australia - Biography.
4. City planning - South Australia - Roxby Downs - History.
5. New towns - South Australia - Roxby Downs - History.
6. Civil engineers - Western Australia - Biography.
7. Roxby Downs (S. Aust.) - History. I. Title.

994.23806092

Published by the Author and Bookends Books
of 136 Unley Road, Unley, South Australia 5063
Phone/Fax: 61 + 8 82710050

Distributed by Calypso Press Distributors
248 The Parade, Norwood, South Australia 5067
Phone 08 8364 1411   Fax: 08 8212 1961

E-mail: bookends.bookshop@adelaide.on.net

Printed and bound by Hyde Park Press
Adelaide, South Australia

Contents

Foreword

The great mineral developments in the 1960s, 1970s and 1980s lifted Australia to a new level of prosperity. New technologies and new industries now capture the headlines and the imagination of the public, but the minerals and other resources industries continue to underpin our living standard. The proof of this is that the value of the Australian dollar fluctuates in close accord with commodity prices.

John Showers has recorded his involvement in some of these historic events, partly in Western Australia but most importantly in the far north of South Australia where Western Mining Corporation (now WMC Limited) discovered the massive copper-uranium-gold-silver orebody at Olympic Dam on Roxby Downs station in 1975. The Olympic Dam operation today includes the largest underground mine in Australia and one of the most extensive metallurgical plants in the world, supported by a town of some 4,000 inhabitants, designed and built as a part of the total development.

John's role in this was to design and supervise the provision of what is today called unattractively the "infrastructure" – the township with its community facilities and many of the service and supporting facilities which make up a modern minerals operation. Starting with just an arid patch of inland Australia, he and his colleagues created an attractive and thriving population centre which made the development and ongoing operation of one of the world's great minerals operations possible.

John's knowledge of this part of Australia's outback goes back more than half a century. The account of his time as an Army Surveyor and subsequent return to the area many years later is fascinating because of the vivid description of the events, the conditions, and the colourful characters. It is impossible not to be impressed by his deep understanding of what most of us would think of as the harsh, arid, and unfriendly inland. His respect and love for this environment comes through at all stages of the story, although in a restrained way one would expect of an engineer.

John Showers has rendered a great service to us all by recording, meticulously and in great detail, an intensely human account of an important facet of Australia's recent history. I am delighted to have been invited to provide the foreword and commend the book to readers.

(Arvi Parbo)

Author's Note and Acknowledgments

At the time it was a great disappointment to be posted to the Army Survey Corps in 1946 and eventually on to the small unit dispatched to spend a year in the arid and then remote north of South Australia. I had expected to be off to Japan in the immediate post-war occupation forces and not in the vacant expanses of the outback.

In recalling the 'highlights' of that year it doesn't sound too bad but there was a lot of fairly ordinary times in between. It was made bearable by the company of those in the unit and those we occasionally met throughout the region. Our Sergeant, Len Beadell in later years went on to become a living legend for his opening up of the interior of the continent.

On leaving the unit and returning to civilian life I did not want to, or expect to ever return to that part of the country again. I settled back contentedly in the well watered mountain areas of Victoria. Many years later a major career change to the mining industry in the Western Australian Goldfields rekindled my latent and unrecognised affinity with the outback. Not only that, it put me in the position to return to the very areas we had mapped and play a part in the major development of the Olympic Dam Mine on Roxby Downs Station.

I have briefly covered the few events of 1947 that remain in the memory and the years following that equipped me for the very wide range of tasks dealt with on the Olympic Dam Project. However the main reason for putting this book together is to provide some record of the events and people that led to the initial production from the Olympic Dam Mine and the establishment of the Roxby Downs Township.

A great number of people were involved in the activities at Olympic Dam during the years covered in this book. The passage of time, reliance on memory, and limitations on coverage has meant the ommission of many notable names that were more than worthy of a mention. This is particularly so in respect to the initial discovery, definition and assessment of the Olympic Dam mineralisation, the mining and metallurgical development, and the specialist engineering involved in these. My participation was largely confined to providing the surface infrastructure facilities whereas the reason for being there was deep underground.

My sparse personal records also proved inadequate and I have relied heavily on information obtained from various publications, reports and other material issued by WMC over the years. A number of the photographs and illustrations are also reproduced from Company records and slides.

I did not start compiling this book for seven years after my last association with Olympic Dam and doubt if it would have seen the light of day without the encouragement and help of many kind people. Firstly Anne Beadell, then Dr Richard Barnes who read the early chapters and suggested I continue.

Later and in particular Frank Boulton who is still with WMC and who was kind enough to provide imformation on the early happenings and residents. He also scoured the emerging manuscript several times correcting my many spelling errors. Terry Dwyer for help with photos and encouragement, Sir Arvi who took the time to read the manuscript and write the foreword and finally my wife Jo for continual corrections and advice.

The events covered in my book only extend to 1991 when I finally retired and returned to the mountains. It is pleasing to note the major expansions in the following years and that the town in the dunes is now home to a population of 4000 and is built to last well into the next century. It is noted for the high level of average income of the residents, low unemployment, and very high birthrate.

It is equally pleasing to note that the original settlement at Olympic Dam has not lost it's identity. The Original camp site and airstrip selected in such a rush years ago, with upgrading still serve the mining project and remain the centre for heavy industry and construction support.

INTRODUCTION

The Olympic Dam Mining Development started with a hole drilled deep into the earth's crust in the driest part of the driest State of Australia in a search for rock containing copper. There was nothing on the surface indicating where to drill and it was known that if a discovery was made it would be deep down. The reasons for selecting this particular isolated location 520 kilometres north of Adelaide was cloaked in geological secrecy at the time. It was part of an innovative exploration program testing for mineralisation under deep layers of barren sedimentary rock. Western Mining Corporation (WMC) had been exploring for copper in South Australia since 1961. In 1975 they were granted exploration licences covering 15,000 square kilometres to the north of Port Augusta.

They selected a likely site to test their theories in amongst the red sand dunes on the Roxby Downs pastoral lease 30 km to the west of the Andamooka Opal Fields. The pastoral homestead lay 40 km to the south west and that of Purple Downs a similar distance south. The nearest habitation to the north was over 100 km distant. The first hole could have been drilled anywhere in a radius of a few kilometres from the only significant surface feature, a stock watering dam. This had been excavated in a small claypan at the time of the 1956 Olympic Games and had been named "Olympic Dam." The muddy water collected in the excavation during the infrequent rainstorms sustained stock grazing the sparse vegetation in this otherwise waterless region.

Whether by design, chance or the proximity of some sparse shade for the driller's caravan, drill hole RD1 was spudded in among the saltbush a short distance from the dam. It was a very good choice of location as after passing through 350 metres of barren limestone,quartzite and shale it intersected 38 metres of copper bearing rock. It was of low grade but sufficiently encouraging to press on and drill further holes at spacings of 800 metres. The next eight holes did not encounter any significant mineralisation. The costs of drilling to such depths in this remote area must have made it a difficult and brave decision to continue. Hole number RD10 was drilled and if it had also been unsuccessful may well have been the last to be drilled at Olympic Dam and possibly in the region.

At a similar depth to the first hole, RD10 intersected a thickness of 170 metres of mineralisation containing 2.1 % copper. This led to an intensified drilling effort, and the definition of an area of 6 km by 3 km containing not only copper but also uranium, gold and silver.

The magnitude of the discovery and the extremely high cost of evaluating its potential and developing a mine led WMC to seek a Joint Venture partner.

In mid 1979 the BP group of companies signed up to provide finance for a 49% interest in the discovery.

The site of all this drilling suffered from an identity crisis almost from the start. The early company advices to the press and stock exchange of the discovery described it as west of "Andamooka". This being the only readily identifiable spot within 100 km on the largely empty road maps of the region.

The further information that it was on the "Roxby Downs" pastoral lease on its own would not help locate it as it was a small spot on the 2000 square kilometre property.

It was when "Olympic Dam" was used in the description of the locality that it became more precise.

The news media seemed to like the "Roxby Downs" title which was soon shortened to "Roxby" in the glowing articles of the brilliance and potential of the discovery.

WMC slowly promoted the use of "Olympic Dam" for the mineralisation, the continued drilling, and the subsequent studies and project development. However in 1979 they kept the name of "Roxby" alive when they called their subsidiary company to manage the project "Roxby Management Services."

I was working with WMC from their Perth office when the news of the success of the first drill hole was announced. I had joined them in 1968 to manage the development of the new Kambalda townships near Kalgoorlie in Western Australia. They had not had the need for Civil Engineers on their staff until it became mandatory for them to provide the infrastructure for this new project. This led to the wider role of looking after all manner of civil works and buildings both at Kambalda and nearby gold operations.

Five years later I was loaned to Poseidon Ltd the legendary "nickel boom" company. When the dust had settled after the boom WMC had entered a "Joint Venture" with them to develop their nickel deposit at Windarra in the far outback of W.A. I had the task of looking after the redevelopment of the largely abandoned old gold mining town of Laverton. This lasted two years with weekly light aircraft flights to site and an introduction to the complexities of aboriginal affairs.

It was then back with WMC in their Perth office involved in continuing studies and costings of infrastructure for potential new projects. There was also a deal of travelling to the operating mines looking after such things as new housing, offices, water exploration and tailings storages. I was attached to the WMC Engineering Services group who assisted Mine Managers as needed. WMC at that time had a relatively small staff and most had a background from the gold mines at Kalgoorlie or more recently from Kambalda. There was a first name relationship from the top down. It was possible to know nearly all the staff in Perth and some at each of the scattered mining operations. I also had contact with consultants in various fields, and the staff of many government authorities.

Those of us also associated with new projects had followed the progress of the early Olympic Dam drilling with great interest. The internal company "bush telegraph" kept us up to date with the drilling and there was great excitement with the success of hole number RD10.

My interest was more personal as I had spent almost a year helping to map that isolated part of the country about thirty years previously. I had been a very junior member of a ragged Army Survey unit of ten men sent to this part of the arid outback to prepare for the establishment of a rocket range. We were led by Sergeant Len Beadell later to detail his outback roadbuilding and other exploits in a series of books. We were the first to arrive among the many that followed to build and operate the township and facilities at Woomera. Our mapping kept us moving and well clear of the encroaching civilisation at the range head.

We were nomads covering the arid lands to the west and north as far as Coober Pedy. We occasionally visited the opal fields, and the isolated station homesteads including Roxby Downs. I remembered it well as it was on a slight rise on an island in the middle of a shallow lake. The normally waterless land had experienced some very unusual heavy storms in the previous year. This had resulted in many of the normally dry depressions and salt lakes holding vast quantities of water. There was also cause to venture in to the maze of sand dunes to the north and east of the Roxby homestead trying to locate the highest points for survey observations.

It was in March of 1977 when WMC decided on the first preliminary study of the likely costs and returns of a hypothetical mine at Olympic Dam. It was to be known as the "Andamooka Study" and meetings were held to spell out the basic assumptions to be taken into account. These were such things as ore grades, mining and metallurgical methods, and infrastructure requirements. The study team consisted of geologists, mining engineers, metallurgists, and engineers, of each discipline, all with specific tasks. I had the responsibility for defining and estimating infrastructure requirements including access, accommodation, waste disposal and water supply.

This enabled me to make a quick visit to Adelaide to check out the latest mapping available and to make enquiries on the availability of services. I took advantage of the opportunity to catch up with Len Beadell for the first time since our army days in 1947. I was aware of his exploits in the years that followed from his well known writings. I had spotted his first book at the airport bookstall in Brisbane when I was attracted by the photo of a Caterpillar Grader on the cover of a paperback. I was taken by surprise to see the title with Len Beadell as the author. Len had been such a shy and private person and at that time had been unaware of his achievements in opening up so much of the outback.

Lenny joined me for dinner at my Adelaide hotel and we yarned well into the night. We first caught up with what we each knew of the movements of the others who had been in the army unit with us. I learnt of Lenny's purchase of his car and house with garage to leave it in while he travelled the globe. (Both purchases made after a visit to his bank and produced wholly in cash to the amazed salesman and builder from his well worn and dusty leather gladstone bag). Then how he returned from his trip to pick up his car, met Anne and her family who had rented the house, married Anne and now was the very proud father of three children. This we would never have predicted in the army days as Lenny was then totally and absolutely girl-shy. We almost had to manhandle him to line him up beside a waitress at the entrance and under the sign for a Port Augusta hotel. This was so we could get a photo on our return from the year in the desert. It was the only photograph I know of with Len immaculately dressed in clean long army pants, a neat shirt with the three stripes and with a girl close alongside. (The photo was given to Anne some years later as a momento of his only known previous flirtation.)

The study progressed slowly and in June I returned to Adelaide, this time to do the rounds of the various government authorities, house and camp builders, and service providers. They were all eager to help with advice and indicative costs and were interested in the prospect of a mine development at

Exploration Drilling Rigs – 1980

some time in the future. It was also opportune to renew contact with engineers from Kinhill, the Adelaide consultants who I had worked with at Poseidon.

On June 10th I returned to the Woomera area after a lapse of nigh on thirty years. Leaving Adelaide at 7.am. on a weatherbeaten Emu Airlines charter aircraft we flew north first to Mt Gunson, an operating copper mine 140 km north of Port Augusta. Here Dick O'Meara, the study's metallurgist disembarked to spend the day on learned discussions and inspection of the mine plant. We took off again this time to land at Woomera and to meet with John Emmerson, the WMC drilling supervisor for the region. Others to meet us were the Woomera Security officers. They had to ensure my specially obtained authorities to enter their "Prohibited Area" was in order. We were supposed to have the blinds tightly closed on the aircraft until we landed to ensure we didn't see any of their secret works. I did sneak a quick look through the front windscreen and out the side. It had all been bare open gibber to the horizons with a few dusty tracks, a dustier dirt runway, and some survey marks, when last seen in 1947. It was still largely bare gibber to the horizon but the airstrip had been enlarged and sealed, and there were clumps of large buildings. Where we had spent days in the fierce heat taking levels to map for a future town as our last task before ending our year in the desert they had actually built a town. There was no time for more than a quick glance around and it looked as if it had seen better days.

It was then away to the north in John Emmerson's 4WD on the narrow station tracks on the sixty kilometre run to the Roxby Downs Homestead. The old building on a rise in the middle of a now dry swamp was gone. A modern transportable house with the usual station outbuildings sat neatly on high ground. We paid our respects to Tom and Judith Allison who now owned and managed the station before again heading north. The next call was at one of John's drills sited on another 40 km before backing off about 15 km and travelling the same distance east to Olympic Dam.

The drillers had their caravans in amongst a stand of scattered mulgas close to a small dry yellow claypan on which a hole the size of a house had been excavated to collect the runoff from the occasional storms. The dirt from the excavation formed a low mound on which was a tank, a stumpy windmill and stock trough. This was the now internationally famous "Olympic Dam."

There was time for a quick look at the country just to the north where most of the drilling had been carried out. Low stable sand dunes covered with mulga and desert shrubs crossed the track at irregular intervals. The interdunal corridors or swales had a light cover of saltbush and grasses. It was then the long drive back to Woomera noting the suitability of the different areas for the access and facilities that would be needed if a mine eventuated.

Our pilot soon had us winging our way back to Adelaide with a stop at Mt Gunson to collect Dick. We returned to Perth the same night which illustrated how much means of transport had changed since my time in the army. In 1947 it took us two days of rough ride to travel from Adelaide to Woomera.

The "Andamooka Study" continued on to completion later in the year initially co-ordinated by Keith Folwell, then by John Oliver, both mining engineers and both having served time at Kambalda. We compiled details of the works, services and manpower that would be required for both an underground and an open cut mining operation. Then we estimated the capital and operating costs of the various components for the years of development and production. It was all based on lots of assumptions and obviously gave very approximate figures for the rates of production adopted.

The depth of the ore made the quantity of waste rock for an open pit option look frightening. Details of the estimated costs and mine production year by year were eventually fed to the analysts to check out. The results must have been sufficiently encouraging for the company to decide to keep on drilling instead of walking away.

I had involved Cecil Forbes, a principal of Australian Groundwater Consultants to help in the the study. Without a reliable and sufficient supply of water for all purposes there could be no future. Cec had found supplies of underground water for us in arid areas of Western Australia when others had failed. The nearest flowing stream at Morgan on the Murray was about 500 km distant and a source of last resort. The recorded drilling throughout the region by pastoralists and miners had only indicated limited amounts of saline water. The coal mining town of Leigh Creek far to the east across the dry salt expanse of Lake Torrens had a large dam collecting run-off from the Flinders Ranges. This did not appeal to us as a source as it was distant, would need a very large dam and we could wait years for the first filling. Cec after detailed investigations pointed us in the direction of the Great Artesian Basin which was known to extend into the Lake Eyre region 100 km to the north. We assumed supply from there in the study and it later proved to be the correct choice for future supplies.

All in all, 1977 was an extremely busy year for me. I was involved with the Environment Impact Statement being prepared for a possible uranium mine at Yeelirrie in WA, drilling for additional water at Windarra and other sundry works. There was even a pleasant week on Christmas Island in the Indian Ocean advising on workforce accommodation.

In contrast 1978 was the opposite with a major downturn in metal prices and demand. There were days with little or nothing to do and finally in November I resigned from my position with W.M.C. We sold our Perth home and returned across Australia to the high country of north eastern Victoria. We had a long term interest in a small family farm in a narrow green valley and decided to settle there in semi-retirement. It was a peaceful time living in a caravan under shady trees alongside a small clear trout stream with a builder organised to build us a house.

The peace was interrupted when a Melbourne based mining company asked if I could advise them on the accommodation needs for a mine in Tasmania. Pleased at the prospect of doing a limited amount of consulting and a first visit to the island state I accepted. It only involved a pleasant trip to Tasmania and a few meetings in Melbourne so did not unduly interfere with the rural lifestyle. The harvest of the walnuts from trees we had planted on the property many years previously went smoothly in April and into May.

Then W.M.C. had me travel over the mountains to spend a day or two looking at possible access routes to a mining prospect deep in a rocky gorge on the Tambo River. This was followed by meetings in Melbourne and ten days back in Perth office in June. All went quiet and we moved into the warm newly built house. There was time to enjoy the clear sunny days that followed the heavy morning frosts. The evenly spaced trees were bare of leaves. Rabbits, the odd fox and an aged wombat with the mange even ventured out from the forrested hillsides in daylight. I leisurely compiled a report for the local golf club on developing a new course on pine covered dredge tailings. Planting shrubs and fruit trees at the new house, a game of golf, a bit of fishing and a bit of farming filled in the days most pleasantly.

It was not to last. I was working with a chain saw when the message arrived asking if I would please phone John Oliver at the W.M.C. Perth office. John had been the Resident Manager and my boss when I first went to Kambalda. He asked if I could go to Adelaide for three months to organise an airstrip, single quarters and a water supply for an intensive drilling programme at Olympic Dam.

This was my "Return to Roxby Downs" that was to last thirteen years and where I was to play a part in the development of the major Olympic Dam mining project and the new township of Roxby Downs.

What follows are my recollections of the year in that part of the outback in 1947 and my return thirty years later.

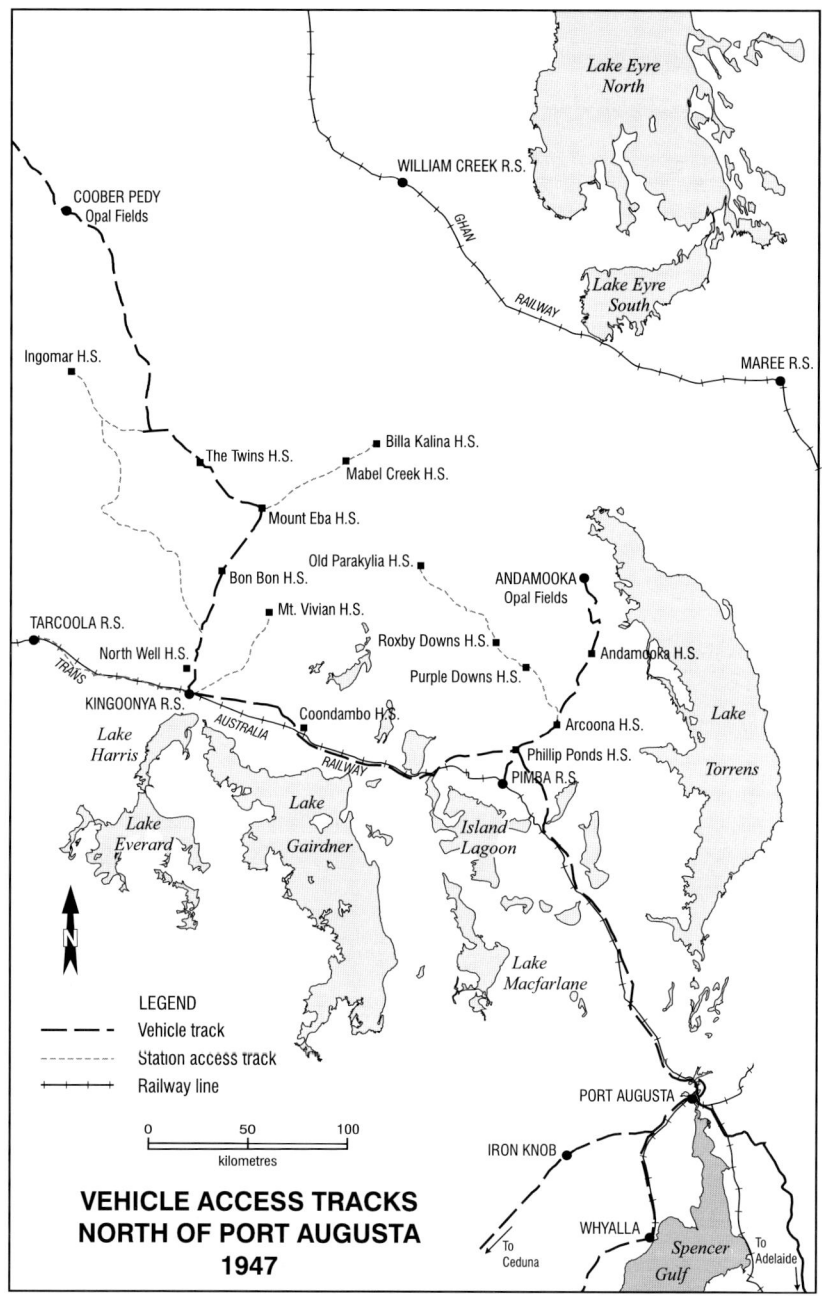

VEHICLE ACCESS TRACKS
NORTH OF PORT AUGUSTA
1947

BOOK ONE

1947 A YEAR IN THE BUSH WITH LEN BEADELL

Chapter 1

PHILLIP PONDS and
A VISIT TO ROXBY DOWNS

The 4WD Army Jeep passed over a shallow stoney ridge where the two wheeltracks continued down a slight grade to disappear into a large flooded depression. On a small island in the middle of the expanse of clear water was the old station homestead. A small wooden dinghy hooked to a long rope was pulled up onto the shallow beach. The rope was fastened to a post on each side of the water and was obviously there for the purpose of pulling the boat across the shallow water.

It was my first visit to Roxby Downs in the North of South Australia. We were in a part of the world that would normally be described as an arid, waterless region sparsely populated with widely scattered sheep stations. This year the larger depressions and normally dry claypans still retained the runoff from unusually heavy rainfall over the previous eighteen months.

Army Camp – Phillip Ponds Homestead – 1947

5th Field Survey Section – Army Survey Corps – 1947

It was early 1947 and our small Army Survey Unit of ten men were in the area to do the mapping of the vast outback area to be used for testing of rockets. We were under the benevolent control of Sgt Len Beadell and camped in tents at Phillip Ponds homestead about 40 miles to the south of Roxby Downs. The old building was originally a coach house on the lonely run to service the few isolated sheep stations to the north.

I was fortunate to be with Lenny on this visit to the island homestead as a courtesy call to let them know that we would be working in the area. This they almost certainly would have already known through the ever reliable bush telegraph as we were resident at Phillip Ponds courtesy of their "neighbours" at Arcoona Station.

At that time road access to this part of the world was along two wheel tracks leading northwards from Port Augusta. The route continued on past Phillip Ponds at the 120 mile mark to Kingoonya, a watering point on the Trans-Australian Railway. There it branched, with one leg continuing along the railway, and the other northwards to Coober Pedy and for the very adventurous, to Alice Springs.

Phillip Ponds was a minor cross roads. A turn to the right led to Arcoona Homestead and on to the Andamooka Opal Fields. A lesser track to the left took you to the railway siding at Pimba at a distance of about 10 miles. Even with all these roads, in early 1947 when we arrived we were lucky to have one vehicle a week passing by.

We arrived at the Ponds on March 12th, 1947 the first of the influx to develop the area for the test firing of rockets and to build the township of Woomera.

The old coach house - homestead was solidly built of local stone on the level ground adjacent to a wide dry stoney creek bed. We were camped in two man tents lined up around the house fence, sharing the area with our three jeeps and 3 ton 'Blitz' truck. A short distance away a small graveyard testified to the isolation and harshness of the country that the earlier occupants had to contend with. Phillip Ponds lay in a wide valley fringed with a steep escarpment to the north. There were stunted shrubs and bushes with the occasional mulga along the valley. The vast tableland beyond the escarpment was a bare, desolate, gibber covered, dry wasteland stretching to the horizon and beyond. This was to be where the future development was to be centred and where our mapping was to start.

Our first task was to establish a network of control points and to this end we were ably assisted by two Survey Corps Majors from Adelaide enjoying a week in the bush. These control points were sighted on the highest ground and each consisted of a concrete marker set in the ground. So their positions could be accurately determined each site had to be visible by theodolite from at least two others, often from a great distance. To this end they were marked by gathering rocks and building a stone cairn or in some cases by erecting a standard trig station beacon. Major Lindsay Lockwood seemed to revel in stripping to the waist and lugging the heaviest of the rocks. Wally Relf, the other Major seemed more comfortable assisting Len Beadell in reading the angles to all the control points.

The two Majors returned to the city after a week, with Major Lockwood transferred interstate. Major Relf stationed in Adelaide remained in overall command of the unit, doing whatever Majors do. The other man of rank in our field party was Corporal Frank Cohen. Lennie was happy to leave all the army paper war and the site organisation of the unit to Frank so he was free to do all the real surveying with his beloved theodolite.

Then came the troops all designated as 'Sappers', the Survey Corps title for those without rank. Mick Waterland, Max Pickering and Ossie Osborne were the driver/mechanics who other than Lennie were the only ones supposed to drive any of the vehicles. Classed as Topographical Surveyors, Harold Watts, Ivan Miller, Bill Fitzgerald and myself were left with the more routine surveying and camp maintenance duties. These ranged from digging the latrine trench, pitching tents, precasting concrete markers, to survey chainman and marking up aerial photographs for map preparation. The tenth and most important member of the unit was George Greenwood our cook. George, a senior citizen, was a broad accented, dour, Northern Englishman who had taken up residence in the house. He soon had all organised in the kitchen where he ruled supreme. The large adjoining room became the mess.

Stores filled another, leaving one as an office for the paper work and the last was fitted out with chairs and a table for use by the troops.

Old Graves at Phillip Ponds.

Supplies and mail came in fortnightly to Pimba on the "Tea and Sugar" train that serviced all the railway maintenance settlements across to Western Australia. We enjoyed a continuous diet of aged mutton from Arcoona Station varied with the odd rabbit which were in plague proportions, some trial roo meat, and if desperate a tin of bully beef. The army very generously provided us with jam! Five large cases - all melon and ginger.

Left to ourselves we progressively gathered all the levels and data in the immediate area which was mailed off for the Army map makers in Bendigo Victoria to start compiling contoured maps. We then started marking up aerial photographs of the areas to the north with all significant features such as fence lines, dams, windmills, tanks, homesteads and station tracks. This we did in two man teams traversing the country in Jeeps with the "Sapper Surveyor" marking the photos and the driver attending to the travel.

The monotony was regularly broken when great clouds of dense smoke erupted from underneath the vehicle or we noticed the floor getting hot. We quickly found the location and cause of the fires. The exhaust pipe under the low slung metal floor would collect large wads of saltbush and grasses which quickly dried out and caught fire. We quickly learnt to skid to a halt, bail out and grab the ever ready shovel. A few minutes of frantic shovelling of gibber, dust and dirt usually doused the flames sufficiently to let us lie on our back and scrape away the charred remains. The distances out from the ponds soon increased and we would camp out in the bush for a few nights each week, particularly as we reached the sand dune country at Roxby and Purple Downs. We worked for six days each week, catching up with our minimum laundry and tent tidying on the Sunday. It was a totally dry camp with the nearest pub well out of reach at Port Augusta. For recreation we would engage in a game of cards on Saturday nights often lasting until daybreak. Lake Richardson a large depression ten miles to the East still held up to six feet of clear brackish water and was a popular spot to have our weekly wash. The Army had let us retain our 303 rifles and also provided a small bore 22. With a plague of rabbits and large mobs of kangaroos about we were soon collecting and drying lots of skins. Subseqently we received pittances for these from the city dealers so we reverted to keeping enough to carpet the tent floors. While we were going about our allotted tasks Lenny was busy night and day with his theodolite reading angles to all the control points, and fixing positions with sun and star observations. He usually had Frank or one of the sappers with him to record his readings as he called them out. This meant waiting on the bare windswept hill into the dead of night until the appropriate star was in the right position. These were the very odd occasions that Len might talk about surveying behind the Japanese lines in New Guinea during the war. When we returned to camp after the nights readings, half frozen and tired Len would sit up in the house office doing the detailed and time consuming calculations often until daylight. Inspired by Lenny's ability I enrolled in a correspondence course in "Astronomical Surveying" to learn how to emulate him. The course papers eventually arrived in our mail bag on the "Tea and Sugar" train and were in loose leaf form. I was half way through the first lesson sitting outside the tent on a sunny Sunday with the course spread out on a box in front of me. With very little warning a strong "Willy-Willy" whirlwind passed directly over in a swirl of dust. Last seen nearly all of the innumerable sheets of paper that made up the course were spiralling skywards. They were getting higher in the dust and moving away irrecoverably into the desert as the mini tornado continued on its way.

Chapter 2

THE GENERAL PAYS US A VISIT

Life was becoming routine after about six weeks when Lenny told us that we were to have some important visitors in the form of three high ranking English officers. The party, limited by our sparse accommodation, would consist of General Evatts accompanied by a Brigadier, a Major, and a driver.

At that stage being well removed from the regular army establishments our dress standards had slipped considerably. We had tended to adopt the very unorthodox example that Lenny had displayed even in Adelaide with it's very pronounced bush flavour. Socks were out both to avoid collecting burrs and to save washing. Shirts with or without sleeves were worn (most times), mainly to carry our ready rubbed tobacco and papers in the top pocket. Shorts were usually worn while the mild weather persisted. All clothing took on a bright red discoloration with the persistent bulldust. The good quality rain water from the house roof was collected and stored in an underground cement lined stone tank. This was carefully conserved for drinking and limited kitchen use. Water for a morning wash and shave was collected in 44 gallon drums from a nearby stock dam. These we lifted on to a timber stand where we could collect a small dish full from a tap screwed into the bung hole. Our standards of personal hygiene tended to be a little primitive. We did have a smallish tin tub but it was a marathon exercise to heat up sufficient water to cover the bottom and once sitting in it with knees under the chin there was barely room to move. It was totally an outdoor exercise which also tended to draw a scoffing but very small crowd. It was far easier to have a swim when the occasion presented itself and accept our soiled state in between times. Hair was a real problem. The poor quality water tended to combine with any soap used to bind it together into a tangled mat. This was compounded by travelling in the open Jeeps with the fierce wind blowing it in to even worse tangles. Lenny had overcome the problem by wearing it at a length no longer than a fraction of an inch. He wielded a mean pair of hair clippers so with the General coming we faced the inevitable and all lined up for Lenny to give us a greatly feared "survey cut."

In final preparation we dug into our wardrobes which were in the form of steel ammunition crates. To our dismay the fine red bulldust had penetrated the supposedly inpenetrable lids and covered all our "formal wear" with a a mottled film. The areas around the tents and house were cleared of accumulated cigarette butts and other foreign matter and the vehicles parked in a neat line. Two more tents were pitched with great

*Len Beadell and
Max Pickering on the
Gibber.*

*Frank Cohen and
Lenny on the road to
Alice Springs*

difficulty in a roaring gale and eventually all was in readiness to receive our
very distinguished guests.

Our role as Sappers during the visit was to generally go about our business
and keep out of the way. The only official tasks were for two of us to wait table
at each meal and only speak if spoken to.

The visitors eventually arrived after the long road trip from Adelaide and
Port Augusta, their vehicle quite a bit the worse for wear. They settled in to
our best accommodation provided by making a couple of the rooms in the
house available. They were early to bed after a lavish meal prepared by George
slaving over his hot stove.

Next day, Lenny escorted them around, and around, and around the
expansive gibber and gullies to the north, no doubt eying off suitable spots to
launch rockets into space and on to Western Australia. Selecting potential sites
for their workshops and a town were also no doubt high on the agenda.

Tired after their long day they settled in for a well presented meal, again
ably prepared by George on his wood stove in his bush kitchen. The two of us
waiting table were quick to notice that they were breaking the "no alcohol" rule

*First Plane
into Woomera*

*Main Street
– Andamooka
– 1947*

of the camp. We thought it best not to mention their offence. The guests were also receiving the close attention of a number of our resident mosquitoes; quite magnificent specimens that had been feeding regularly since our arrival. We had become quite adept at grabbing them out of the air as they flew past. This we now did unthinkingly as we attended the table much to the great admiration of the General. I have no doubt that on his return to the Regimental Mess in England many after dinner tales were told of the strange ways of the colonials.

Not relishing the prospect of future arduous road trips the instruction was left for Lenny to pick a spot and peg out a site for an airstrip. It was some months before this was ready for use as in early May the skies opened up and over three inches of rain fell in quick time. The gully beside the homestead became a raging torrent and cut us off from the outside world. Worse still it was impossible to travel by vehicle in any direction without getting bogged every few yards. We spent day after day digging the Jeeps out of thick gluggy mud and really getting nowhere.

At the end of a fortnight the country had dried out a lot and we were only bogging down two or three times each day.

There was worse to come!

THE MOSQUITOES ARRIVED IN PLAGUE PROPORTIONS!

They arrived almost instantaneously in their billions and trillions. Instead of the odd half dozen or so that had settled permanently in each room of the house and under each mosquito net they were in swarms. They were also totally active throughout the day as well as at night, so there was no escape. Travelling at speed in an open jeep produced enough breeze to keep them at bay. However as soon as the vehicle stopped or slowed down, no matter where, you were surrounded. A bare arm, wrist, or face was literally black with the feeding hordes within a minute. Fortunately the weather had cooled down considerably since the rain so we were able to clothe ourselves from neck to knee. We had no insect repellent so tried everything from motor grease to hair oil, all with limited benefit. It was nigh impossible to use a theodolite or to mark up aerial photographs without them being anointed with squashed mosquito. Getting bogged became an ordeal. We even resorted to bumping along the sleepers of the Trans-continental railway if travelling in that direction rather than risk the muddy ground at the side.

By some miracle about ten days after their arrival they all disappeared as suddenly as they had first appeared. Whether it was the length of their life span, they migrated, or were blown away we could only surmise. Our hardier and larger resident mosquitoes once again had us to themselves.

Our exclusive occupancy of "The Ponds" was drawing to a close. Two civilian surveyors and a heavy road grader with operator were the first to arrive. A few Air Force types soon followed and Lennie had to set out a line of pegs for the airstrip runway on the gibber tableland so they would feel at home. The grader was soon clearing the stones and filling the bogholes along the line. The word was passed to Adelaide that the dirt strip was cleared and return signals told us to expect the first plane. There was great excitement in mid June as we all gathered at the end of the long runway looking expectantly at the clear sky in the south. Right on time, an Airforce Dakota was overhead checking us out with a couple of circuits, and noting which way the dust was blowing. Shortly after the pilot lined up with the narrow strip of bare dirt and brought the machine in to a perfect landing. The pall of dust that was raised along its path was nothing to the blanket that enveloped the waiting crowd as it turned to park.

The passengers were loaded on to the Jeeps for a tour around the gibber, ate their cut lunches and reboarded the aircraft to return to the city. The flight was very well timed for Frank who had become quite ill over the previous few days. Approval was given to load him out to hospital in Adelaide on the return flight. Despite our best efforts to demonstrate signs of serious ailments, us Sappers were refused evacuation and the chance of a few days in the city.

The General now had his airstrip.

Chapter 3

OUR LAST MONTHS AT THE PONDS

had brought my cheap folding camera with me when we came to the Ponds and kept taking odd shots of the local scenery and landmarks to record our stay. Among these were a series showing the progressive arrival of the first plane from approach to touchdown and its dusty progress along the strip. The ultra security now coming into force prohibited new arrivals from bringing in a camera. Harold Watts, forever on the lookout, recognized a golden opportunity to add to our supply of future "beer money."

Gathering some primitive equipment, some hastily ordered chemicals and the best rain water we went into production. Night after night we toiled in the dark developing film and making prints with the aid of a kerosene Tilley light.

The quality was a bit suspect and other than the creek in flood, the first aircraft, a few sandhills and mulga the scenery wasn't outstanding. Nevertheless they sold like hot cakes to the increasing population who had to find something to pad their letters home. They also must have had a reasonable keeping quality for years later a number were reproduced in one of Lenny's books.

Our work was now taking us further afield and we spent days to the north and north west in the sand dunes searching for the trig stations the first surveyors had established throughout the region in 1875/76. Len wished to tie these into his new triangulation survey as a basis to extend the mapping. Once located he would set up his theodolite on or adjacent to the standard markings and read the angles to all other trigs he could sight. We knew generally where they were from the detailed field notes and calculations the old timers had recorded and left as a permanent record.

The aerial photos we carried were a help in locating the general area but the small scale made them difficult to follow in the tangled scrub that covered the dunes. The surface was still moist and reasonably firm from the rains but the steep sandy slopes were as much as the Jeep could handle when we tackled the larger dunes. Knowing that the very highest point in each location would have been selected we thought that by topping a dune and scanning the full horizon we would quickly locate the highest ground. This was not often so, due to it being difficult to get a clear line of sight above the high bushes that seemed to thrive on the tops. Also it always appeared to be a fraction higher a bit further on. Our persistence was eventually rewarded when one was found just to the west of Shell Lagoon on Purple Downs Station. The claypans and hollows around the Station homestead were still holding water from the rains but the house was high and dry unlike their neighbours at Roxby Downs.

Some time later we found another of the old trig stations. This time it was visible from a distance as it stood at least six feet high and as much in diameter. Being deep in the dunes,remote from any rock outcrops it was built entirely of mulga logs. Although over 70 years had passed since it had been so meticulously and symmetrically assembled it was in perfect condition. The dry climate had so preserved the very dense hard wood that the axe marks still showed clearly. It made us realize what difficulties these early pioneering surveyors and their chainmen must have faced as they established this first network of trig stations far from the nearest settlement on the coast.

Having worked almost continuously for some months Len thought we should have a break. Some time previously we had helped out some residents of Andamooka whose vehicle had broken down when returning from a trip to the city. They had extended an open invitation for us to visit them on the opal fields. We set off on a Friday afternoon in two Jeeps taking about three hours on the unmade track to cover the seventy odd miles arriving just on dark. We were received warmly and after a shared meal we spent the evening yarning and admiring the magnificent collection of opals our hosts had collected. We could barely wait for daylight to go out and look for some for ourselves. We all had our swags so it was early to bed and early to rise.

In the daylight the magnitude of the diggings became apparent. There were shafts in all directions each with their accompanying heaps of white dirt and rock that overlays the seams that contain the opal. After a conducted tour of several shafts and the area we were given permission to noodle. We quickly took to the waste heaps hoping to find colours the miners had missed. Our luck was variable but we all found keepable specimens without making our fortunes and ended up buying some nice polished stones from the wide range available. They would be much appreciated when we returned to civilisation and we could claim that they came directly from the mines.

Then it was back to the Ponds and the weekly duties. Our unit numbers had by now declined with Ivan Miller leaving for discharge. George, our cook had been evacuated to Adelaide with a badly lacerated arm after being thrown around in a Jeep in a night accident. This was really bad because the troops then had to take it in turns to be the cook for a week at a time. The influx of workers arriving daily meant a loss of our privacy at the Ponds so we were pleased when Lennie decided it was time to move. He had arranged for us to relocate to Coondambo, sixty miles further west along the "Trans" line.

Chapter 4

TO COONDAMBO AND COOBER PEDY

Decamping and loading the 3 ton truck with our meagre possessions was quickly accomplished. We then caught and loaded into the Jeeps, "Biddie" the dog, six cats and the joey kangaroo we had managed to collect during our stay. Travelling in convoy the only mishaps were bogging the truck and a flat tyre or two before we reached our new camp. Lenny had negotiated for us to occupy the shearer's quarters at the East Well outstation on the Coondambo pastoral property. It was sheer luxury with rooms to ourselves, a roof over our heads and a real floor. This lasted about six weeks while we marked up all the photos of the country within reasonable travelling distance.

There was a large kitchen and with each of the rostered cooks on a slow learning curve, those of us that were able tended to stay in the bush for a few nights each week. We were happy to settle for char grilled kangaroo nicely burnt to our particular tastes on a mulga campfire rather than chance "the chef's surprise" in the camp.

At weekends we fitted in well with the station activities and got to know Joe the station overseer and the aboriginal stockmen who also were resident at East Well. There was humour and lots of advice thrown at us when we helped them to yard the sheep and tail and mark the new lambs. Lenny was also in his element surrounded by the aboriginal children instructing them on the use of the theodolite. Their laughter and shrieks on seeing everything upside down when they were allowed to look through the telescope could be heard at a distance.

We also had the opportunity to ride the magnificent horses they ran on the station. Ossie, brought up in the inner suburbs of Sydney took to riding and station activities with great enthusiasm. In later years on leaving the army he settled on a nearby property and shortly after became the manager.

The next move was to Mt Eba Homestead where again we could occupy part of the the shearers quarters. This was about 70 miles to the north, but a travel distance of 110 miles on the unformed track to Coober Pedy via the railway settlement of Kingoonya.

Once again the 3 ton truck was loaded but this time there was no kangaroo as he had left us and returned to bush living. The cats had been gladly accepted at East Well so we only had the dog to take. We left Coondambo about mid morning. The winter sun was well up but lacking warmth so we were rugged up in woollens and greatcoats in the open Jeeps.

Aboriginal Stockmen with Joe Stafford at East Well – 1947

The 3 ton truck headed the convoy driven by Mick, He had as his passenger George, the cook, now recuperated and back from Adelaide much to the great relief of all. The three jeeps followed at the slow pace set by the loaded truck. Lenny with all his survey gear barely had room for Bill as his passenger. There were three of us in each of the other two vehicles as our numbers has increased by one with the arrival of a new recruit. He was a tall affable red headed freckled Queenslander known only as Bluey. It was pleasant travelling along the narrow dirt track. The favourable season meant that almost every bush, shrub and mulga was covered with new growth. Here and there were clumps of brightly coloured wild flowers and sometimes the vivid scarlet of the Sturt Peas.

By noon we had reached the junction at Kingoonya, a watering point for the steam locomotives on the Trans-Australian railway to Western Australia. This was where we were to turn to the north, leaving the rail line and head in the general direction of Coober Pedy and Alice Springs until we reached the sheep station homestead at Mt Eba.

Lenny decided that he would continue on in his Jeep with the slower truck while the rest of us in the other two Jeeps could stop for a leisurely lunch. Looking around for a suitable spot we happened to see the settlement of Kingoonya on the far side of the railway line. Venturing over, we passed the large overhead water tanks, the galvanised iron sheds and workshop, and the short line of identical weatherboard railway houses. In another line across a wide stretch of hard bare dirt was the Kingoonya Hotel, a store and a hall.

It would have been unsociable not to call in and make ourselves known so we parked at the hotel and went in and met the genial host Mr Jack Crosby. We also met his lovely wife and his three equally lovely daughters. It was totally too much of a temptation for us! We had not enjoyed a beer, nor seen a girl even in the distance, for over six months. We settled in for a quiet beer

Kingoonya Hotel

and a sandwich for lunch in preference to a tin of bully beef and boiling our quart pots for a cup of tea in the bush.

It was absolutely amazing how the time passed! In no time at all it was well after three and it occurred to us that perhaps it was time to get under way again. This we eventually did after some discussion and making tentative arrangements for a return visit. There were suggestions of a game of cricket with a local team if they could muster one. Equipped for the remainder of the trip with the odd bottle or two, and one for George who had gone ahead with the truck to have the evening meal ready on our arrival, we finally got under way.

Darkness had already well and truly set in when we eventually arrived at the Mt Eba settlement in high spirits.

A very, very cool reception awaited us which probably was not that all surprising in view of our expected time of arrival of about three o'clock. None of us had found it necessary to stand to attention since arriving in South Australia at the beginning of the year until that night. Lenny, uncharacteristically in full Sergeant's mode soon had us rigidly in line and very subdued. With Bob Crombie, the manager of the station, as an interested observer, he gave us a dressing down we were not likely to forget. Len had never been known to swear or use intemperate language of any sort. Neither did he smoke or drink, and was normally silently tolerant of those of us that did. Dinner was out of the question as George had let it go cold, so we slunk off to our quarters in disgrace. In retrospect it was obvious that we had been dealt with rather leniently although we could claim that there had been "extenuating circumstances". On the other hand we had been in a position of strength as we were two thirds of the unit. There was no place to lock us up and it would bring the mapping to a standstill had we all been shipped off to the Army brig in Adelaide. (via Kingoonya.)

We made up for our transgressions spending the next weeks marking up the aerial photos with new vigour and covered a wide area radiating out from Mt Eba. The country was more harsh than at East well with less mulga and large expanses of hard baked soil with a light cover of stones. In addition to the Station Homestead there were numerous outbuildings that go with an isolated and self contained sheep property. There was also an airstrip which was a scheduled refueling stop for the Douglas DC.3. aircraft used on the Adelaide to Alice Springs and Darwin air link at that time.

The air traffic was controlled and monitored by two resident wireless operators who shared the shearer's quarters with us. They operated at a well equipped radio hut situated alongside the airstrip. This meant we could sometimes join them for the novelty of listening to a radio program on their powerful short wave receiver. The only radio we had with the unit was a small battery operated unit for Lenny to pick up the Greenwich time signals. These were essential when carrying out his astronomical observations in the dead of night and at noon.

We all crowded in to the airstrip radio hut on the night of the "Royal Wedding" in London and probably had better reception than those in the city.

Max and I would have appreciated a means of communication shortly after our arrival at Mt Eba when stranded half way back from Coober Pedy for several days. We had set off from Mt Eba in the 3 ton truck loaded with 44 gallon drums of petrol to be stored in a dump at the opal mining settlement for Lenny to pick up and use on his first exploratory expedition to the west along the centre line of the proposed rocket range. The 85 miles up the ungraded track on the forward trip were uneventful and we arrived early enough to have a good look around the field before settling in for the night. The only above ground buildings were two tin sheds serving as a "post office and store. The miners residences had all been dug into the face of the central hill. The soft stable rock could be easily excavated to form neat, dust free and cool accommodation. Shelves and wardrobe recesses could be readily carved into the walls and in some there might even be a thin seam of opal displayed.

We spent the night underground in one of the vacant units after a pleasant evening getting to know some of the miners and purchasing small samples of some of their polished opals.

Early next morning on the way back to Mt Eba our troubles started with a blowout and a ruined tyre as a result. The spare wheel was soon fitted to get us under way again but a mile or two further on the engine stopped and wouldn't start again. This we found was due to a faulty and unfixable fuel pump. Looking in the tool kit we located a length of thin rubber tubing and found that we could use it to siphon fuel from a 4 gallon drum directly into the engine's carburetor. This got us going again, but very slowly as I had to sit precariously on the truck roof and balance the fuel drum while Max drove. The day was clear and hot and we were pleased to be making our way on the rough track until a loud bang signalled another blowout. Again there was a large gaping hole in the tyre which was beyond repair so there was no hope of further progress. We were still about 50 miles from home, in the midst of an unbroken expanse of open gibber, with the odd clump of stunted saltbush, and not a tree in sight.

The only traffic that could be expected on the road was that of old Jake on his fortnightly mail and supply run from Kingoonya to Coober Pedy. He wasn't due for another week, it was too far to walk, so we would have to sit and wait. Surely someone at the camp would notice that we hadn't arrived back and come looking for us. Two long, hungry, uncomfortable days later they did.

Chapter 5

BACK TO KINGOONYA

We must have worked well in the first weeks at Mt Eba and as a reward those that wished were given approval to spend every second weekend in Kingoonya. The telephone links at the time were a few party lines connecting homesteads, usually using the top wire of fences as the conductor. They only worked in dry weather, were subjected to all sorts of breakdowns, and were repaired and maintained by the station people when they had the time. In spite of the unreliability of this means of communication we managed to get a message to the hotel of our intent to visit and so a cricket match was organised.

Friday afternoon saw six of us lined up for the rough and dusty trip, particularly for the four travelling under a canopy on the back of the old 3 ton "blitz" truck. We soon freshened up after a welcome shower at the hotel. Accommodation was in a large dormitory style bunkhouse in a tin shed at the back of the hotel. This was normally used to provide for the overflow accommodation at those times when there was a big event in town such as a race meeting.

On the Saturday the numbers in town were swollen by an influx of station people, most on their regular trip to collect mail and supplies. Others no doubt came to eye off us newcomers, who would be traversing their properties in the coming months. All wanted to see if there was anything to be learnt about this "rocket range thing" and how it might affect them.

In the afternoon there enough about to make up two sides for a game of cricket and somehow sufficient gear was found to get the game started. The

Mount Eba Homestead

At the Kingoonya Hotel – Sunday morning

"oval" was the hard red dirt that formed the overwidth road between the pub and the railway which were deemed to be boundaries. There were no limits in the other directions so there was a lot of running as the bare outfields were very fast. Any shot that reached or neared the pub verandah was the cause for an interruption to play, for all to enjoy drinks. Needless to say, the standard of play deteriorated as the day wore on but the game was enjoyed by all including the vocal audience in the shade of the pub verandah.

That evening for the first time in many months we had the pleasure of sitting up to a table with a cloth on it and having a home cooked meal served up to us in the hotel dining room. The three daughters from the hotel had another lass from Mt Vivian Station staying with them for the weekend. We had the added pleasure of their company at a social gathering at the village hall that night which topped off a memorable day. Sunday was a day of quietly resting with a game of cards on the pub verandah before heading back to Mt Eba in mid afternoon.

We must have passed muster with the locals and the station folk for we were accepted warmly into their community during the rest of our stay in the region.

The work for those of us marking up the aerial photographs was now taking us further and further out from Mt Eba. Ossie and I were regularly working together and would camp out in the bush for four or five nights at a time. We would return to camp to replenish our supplies and for a good meal or two at weekends. On returning one Friday night we learnt that we had missed out on the exciting happenings of the week. Apparently the Douglas DC3 airliner on it's regular run to "The Alice" had made an emergency landing when one of engines failed. This stranded it and its load of passengers and crew at Mt Eba for the night which really filled up the homestead and shearer's quarters. George had to work overtime in the kitchen with the help of the air hostesses and eagerly volunteering sappers. They managed to feed the influx with the meagre supplies available. A replacement aircraft carrying a spare engine and mechanics to effect the changeover arrived from Adelaide next day. The engine was unloaded, the stranded passengers and their luggage reloaded on the new arrival and off they went. Once the new engine was fitted in place of the broken one it was time for a test flight. All in camp were invited as guest travellers on this free flight to view the desert from a new angle. Meanwhile Ossie and I were slaving away in the remoteness of the bush unaware of what we were missing. We had been allocated the area to the east of which took in part of Miller's Creek Station. It was a barren but spectacular landscape when viewed from the top edge of a long breakaway escarpment about ten miles from Mt Eba. The view must have been almost as good as from the aeroplane and we had the time to absorb it at our leisure. You could see for miles over the expanse below dissected with endless humps and hollows and the headwaters (?) of innumerable dry water courses and gullies. These either joined Millers creek to finish in a landlocked salt lake, or one of the major creeks that fed into Lake Eyre, one hundred miles to the east. It was the first time we had seen such broken country laid out before us and the impression of total isolation was overwhelming. Fortunately our marking up did not require us to go further in this direction and we continued on towards the south and Mt Vivian station. Ossie and I camped at a boundary riders hut at the north end of the property and were fortunate to be invited to the homestead for the occasional meal.

We now looked forward to our next visit to Kingoonya, particularly as we learnt that there was to be a dance on the Saturday night. We all dressed up in freshly washed but unironed clothes. Our hair had grown again over the months since Lenny's ruthless barbering so it was combed and slicked as best the goo from the bore water would allow.

The town population of about twenty had been joined by station people from up to one hundred and fifty miles distant, no doubt welcoming a social outing. The ladies valiantly put up with us as out of practice dancing partners with inappropriate footwear. There were several piano players in the self reliant community to vary the music and lots of conversation between dances. It was a most enjoyable night topped off with a supper of home made cakes. We had all been on our best behaviour and had even taken time off from the dance to put Bluey to bed. He was on his first visit and had cured his thirst a bit too rapidly.

All was quiet a little after midnight. Pat,the lass from Mt Vivian Station and I were still about having volunteered to wash the cups and give the floor a sweep at the hall. We had just returned to the hotel and were sitting in the parlour for a chat remarking how peaceful it was. The silence of the night was suddenly shattered by a piercing shriek. This was followed by kookaburra like raucous cackling at full volume coming from the living area at the rear of the hotel building.

Our first thought was that perhaps Bluey had woken from his deep sleep, and disoriented in strange surroundings had wandered, and blustered into the girls room.

Jumping up, we rushed down the hallway seeking the source of the disturbance, uncertain what lay ahead, but prepared for the worst. There was a light on in a room at the end of the hallway where the noise was coming from. This proved to be Mr and Mrs Crosby's bedroom. Jack was in convulsions, lying on the bed and thrashing about wildly. He was a very big man and we learnt later was reputed to be the past heavyweight champion of the Northern Territory.

Mrs Crosby, a petite lady was desperately trying to hold her husband down so he came to no harm. Yelling to Pat to go and rouse the boys to come urgently I took over from Mrs Crosby in trying to keep Jack steady.

After an age Frank, Ossie, Harold and Max arrived and we managed to hold the situation until Jack settled down with exhaustion. By that time everyone was up and about. Mrs Crosby, now relieved of the immediate worries now took control. Firstly she managed by some means to get a phone call through to I think her brother who was a Medico in Adelaide. Although no doubt woken from a deep sleep he was able to advise the best procedures to follow, to prepare the patient for evacuation to hospital in the city. To bring him to a relaxed state a hot bath was recommended. The bathroom being just across the hall from the bedroom this did not seem to be a problem but it was. The doorway into the bathroom was extremely narrow and we found that it would take four of us to carry Jack to the bath. There was much to-ing and fro-ing with lots of gratuitous advice from the onlookers before we finally had him safely in the hot bath. The treatment had the desired effect and and shortly after we had Jack back in bed resting quietly and calmly.

It was now after 3am and the Trans-continental train was scheduled to pass through Kingoonya on the way to Adelaide at about 5am. Mr and Mrs Crosby were to be put on the train which normally did not make a stop unless passengers were prebooked, or the engine was low on water. Pat, Frank and I volunteered to wake the Stationmaster who lived in a railway house alongside the line and make sure he was up to stop the train. Piling into a Jeep parked outside the hotel we sped across the gravel and screeched to a halt at the back of the darkened house. The night was pitch black as we ran across towards the back door. I didn't make it first go as I couldn't see the clothes prop looming up at head height that collected me squarely and laid me out on my back. Pat helped me to my feet while Frank pounded on the back door of the darkened house. It took repeated pounding and loud calls to rouse the stationmaster who appeared to have had a very heavy night, and wasn't at all keen to be disturbed. On finally getting through to him that Jack Crosby was ill he volunteered the advice from his bedroom to "Go'way—give him a fly in a glass of water."

We persisted with our pleas until he eventually appeared, pyjama clad at the door. Staying until assured that he would take action to stop the train we returned to the hotel. Suitcases had been packed for Mrs and Mr Crosby and transport organised to get them to the train, which stopped as arranged.

The understanding conductors ensured that they were comfortably settled in a vacant sleeping compartment, the train took off for Adelaide, and once again we returned to the hotel. By this time it was nearly dawn, we were wide awake and decided on an early breakfast and lots of coffee. The day was spent relaxing in the sunshine on the pub verandah, yarning, playing cards and enjoying a very quiet time, before the long trek back to Mt Eba.

Mr Crosby remained under treatment in Adelaide for some time and Mrs Crosby stayed down there for a few weeks. In our subsequent visits to Kingoonya, or when passing by, we were pleased to help out at the hotel as much as we could. We were almost accepted as family and as well as lifting and carrying, we gained experience behind the bar. If we arrived at a late hour for a weekend stay we were trusted to open the bar for a quiet drink and move in to the bunk room in the back shed.

Chapter 6

THE BOGGY SALT LAKE

The unit became more mobile with Ossie and I ranging over the wide area to the north east of Kingoonya. Harold was allocated the task of marking up the photographs of the remaining area around Kingoonya as he was growing very fond of Doreen, one of the Crosby daughters. He and Max based themselves in a bush camp just out of the settlement. Lenny had returned from his first expedition out from Coober Pedy and now needed to fix some photo points with sun observations about 60 miles to the west of Kingoonya so I joined him for a few days.

We left Harold's camp very early to travel out to the area knowing from the photos that we would have to wind our way through a maze of sandy country with clay pans and salt lakes. Plotting a course from photo to photo we eventually located the target area. Fortunately we were able to identify sufficient landmarks to pinprick our exact position on the very small scale photo. The theodolite was set up for reading the angle of elevation of the sun at the precise time it reached its highest point in the sky. All went well so the gear was packed and we headed the Jeep for home. A large elongated salt lake barred the direct route but the air photos showed that it narrowed to a width of about 100 yards in the middle. It was worth a look so we headed in that direction. On the way we staked a couple of tyres and also stopped for an encounter with a scrub turkey. This delayed our arrival at the salt lake until later in the afternoon when we were pleased to see that the surface was quite dry with a crust of dazzling white salt.

A quick check on the photos told us that crossing the narrow neck would save a round trip of of over 10 miles on the rough scrub covered dunes that edged the lake. We walked half way across the blinding surface just to make sure it was firm and agreed to chance the crossing. The Jeep hardly dented the surface until just past the half way mark when it broke through the salty crust into the black gluey mud that lay beneath. There it stopped, bogged to the axles refusing to move either backwards or forwards. We were very experienced in jacking up vehicles from bogs and knew that we needed lots of solid packing material under all wheels to prevent them settling back into the mire. We also knew that the solid packing would need to extend to firm ground along the line of evacuation. Our only hope was to gather the dry mulga wood that lay scattered under the scant vegetation that lined the shoreline. Several hours of dragging these and cut branches across the 100 yards of salt in the fading but still hot sunshine led to several failed attempts to free the vehicle. We persisted as darkness came, and exhaustion approached making only a few feet of progress on each abortive try. There was a half moon and the light reflected from the salt surface helped in our efforts to locate more packing.

As midnight approached we elected to take a break and do something about a meal. Lighting a fire with some of the precious mulga sticks we soon had our individual "quart pots" boiling and tea leaves thrown in. Then came the main and only course available - the scrub turkey char grilled on a shovel and eaten with black muddy well salted fingers.

Semi-refreshed we continued with what was now chop, gather and drag to add more body to the corduroy pavement under and behind the Jeep. It was then we discovered that the engine was overheating badly. A torchlight inspection revealed a leak of radiator water from a hose apparently holed by a mulga stick from underneath. A temporary repair made with insulation tape slowed the loss to a dribble. However after topping up the radiator we were down to the last of our precious water.

Finally in the early hours we had success and the mud covered machine was back from where it had left the shore. Before heading for home a hole was dug into the black goo under the salt well out from the shore. This slowly filled with relatively clean saturated brine which we scooped out and used to fill all our water cans. Then we topped up the radiator with the same mix and headed home. This time we took the long way around the lake, picking the best looking route from the photos with Lennie using the familiar stars to maintain the correct direction. It was slow laborious travel across the bumpy ground with frequent stops to add brine to the engine.

Soon the stars faded, the day dawned, and the sun rose into a cloudless sky. It was well into the morning when we finished the slow and uncomfortable cross country travel and picked up a station track leading into Kingoonya. The water cans of brine were nearly empty as we approached the railway crossing leading to Harold's bush camp. The final obstacle then appeared in the form of one of the infrequent freight trains of great length. We had to wait impatiently while it slowly cleared the crossing and reported in to the relief of those who would have had to go and look for us.

The last weekend in October was a time of great excitement in Kingoonya. The annual race meeting was held with the running of the Kingoonya Cup the big event of the day. We arrived early on the Friday night as did what appeared to be every man, woman and child from every station north of Port Augusta. They came in all manner of transport bringing with them a large number of horses including their hopefuls for the big race as well as entrants for the lesser events.

Most of the station people were self contained and set up little camps in the bush just out of town with a yard for the horses and the station dogs in attendance. We were in our usual quarters at the back of the pub. It happened to be my 21st birthday on the Saturday of the races so we purchased a crate of bottled beer on the Friday night. This was wheeled into the bunkroom and it seemed a good idea to have one or two after a pleasant dinner in the hotel dining room. The nett result was that we had a marvellous night but felt very seedy on the Saturday morning.

The racecourse was just out of town and by lunchtime the crowd was building up with the station folk enjoying the opportunity to socialise after their remote lifestyle. As well as the races the day was planned with all manner of other activities on horseback. Ossie, Harold and I had ridden the station horses while camped at East Well so had looked forward to a ride in one of the races but only Ossie made it to the starting line. Harold and I were suffering too much from the previous night's excesses. The race track was smoothed bare dirt, dry and dusty with a rough timber running rail, and a small stand for the judges. The station horses were bred from good stock and the races were keenly contested. Ossie was given a mount in the big race but even with our vocal support was unable to get a place after the field passed the post in a cloud of dust.

At night there was the usual social and dance, this time with a record crowd.

We continued with marking up photographs from the Mt Eba base until the end of November with the temperature soaring and continuous dry weather. It was easy to appreciate how inhospitable the area from Phillip Ponds north to Coober Pedy and beyond must have been to the early settlers. Until the first bores and catchment dams were developed it would have been completely waterless other than for short periods after heavy rains.

A lot of our travelling involved following narrow winding tracks through mulga thickets where visibility was limited. To ensure a cooling air flow, we had the Jeeps wide open at the sides with the canvas roof either up providing shade or folded down and out of the way. There were many near misses as kangaroos, emus and the odd dingo scooted across the track ahead of us. On one occasion an extremely large emu left its run a bit late and hit the open side of the low slung vehicle at speed.

Being long legged and top heavy the momentum was sufficient to catapult the bird into the Jeep just behind Ossie and I. Overcoming the initial shock almost instantaneously we decided that sharing the cramped space with an agitated kicking emu was not for us. We threw ourselves out of the Jeep landing heavily on the bare dirt. The vehicle continued on with its new passenger until it came to a stop and stalled in a clump of mulga. By this time the emu had found a toehold, kicked its way out and continued on its way no worse for the experience. Picking ourselves up and finding no real harm done to the Jeep or to us we proceeded on our travels.

Lennie and Frank had been spending a lot of time over the months with sun and star observations establishing where on the earth's surface identifiable locations on aerial photos were. These together with the positions of trig stations allowed the Survey Corps map makers in Bendigo,to set out the control framework for their maps. They could then add on the physical features such as salt lakes, fence lines, bores, dams, tracks and buildings that we had noted or marked on in our travels. There was sometimes the temptation put a local name of our choosing on some feature on the photos which were meticulously copied onto the finished maps by the far away unknowing draftsmen.

Our next move was away from the friends we had made at Mt Eba and Kingoonya. It was back to the Ponds area which was now a hive of activity. Rather than move in with the crowd we settled for a bush camp out to the north on Purple Downs territory.

The planners had by now picked out sites for a range of installations and facilities including the location for the Woomera township. This meant that detailed plans at a far larger scale were required over the selected areas. We also needed to extend the detailed mapping further to the North into Roxby Downs and towards Andamooka.

Anxious for a return to civilisation before Christmas we set about gathering the additional data required as quickly as possible. The daytime temperatures were constantly above 40 degrees and most of the work entailed levelling on the open gibber with no escape from the blazing sunshine. Lenny would man a theodolite equipped with three horizontal cross hairs. These when set on a levelling staff and recorded with angular readings enabled later plotting of the position and level of each spot where the staff was held. The team consisted of the instrument man, an offsider recording the readings to each staff position, and two or three of us sappers as chainmen. All we had to do was to walk endlessly from position to position holding the staff upright at each spot until a wave from Lennie told us his readings were complete. As the heat shimmer distorted readings from mid morning, we would start at first light, and spend the middle of the day in whatever shade we could find. Starting again in the late afternoon it was after dark when we finally made it back to our camp.

It was in this manner that we took levels over the whole area of the future township of Woomera in the last few weeks leading up to Christmas of 1947. Tramping ten to twelve miles on foot over the bare sloping gibber each day was the final test.

We lost no time in packing our modest possessions and heading south when this last task for the year was done, and we reached Port Augusta in time for an

Lennie – Dressed Up – on return to
'Civilization'

evening meal. Accommodation was at a hotel, courtesy of the army. It was the first time that we had slept in a real bed with sheets since arriving in South Australia. Before we left in the morning we managed to get Lenny to dress up for a photograph at the hotel. He was fitted out in neat long khaki drill pants, a clean shirt with his three stripes showing, and had polished boots on. A second 'exclusive' photo was taken of 'girl shy' Lenny with a hotel waitress close beside him under the prominent 'Saloon Bar' sign.

It was late afternoon when we arrived back at Keswick Barracks in Adelaide. Our accommodation there was the same as it had been during our previous stay early in the year, before we headed north - in the prison compound. Whether this was the only accommodation available or that the brass thought us unfit to stay in the camp proper we didn't learn. In any case it was out of everyones way on the far side of a concrete drain and we had the place to ourselves.

The next day, properly attired for leave we were issued with leave and rail passes and headed off home for our Christmas break. It was a joy to get back to the green grass, forests and clear flowing streams of north eastern Victoria.

Returning to Adelaide in mid January, I found I had been promoted to corporal. However I elected to take my discharge from the army rather than return north with the unit. I had no wish to embark on a career in surveying and had a job to return to in the engineering department of the Electricity Commission in Melbourne.

Woomera, and the arid interior dropped right out of my thinking for the next twenty years, and there seemed little chance of me ever returning to work in that area.

BOOK TWO

1948 TO 1979
LEARNING TO BE AN ENGINEER

Chapter 7

IN THE 'HIGH COUNTRY'

On discharge from the army I resumed employment with the Victorian Electricity Commission and spent a year working in Melbourne. This enabled me to attend night classes one night a week. By the end of the year I had passed the single subject that was outstanding to qualify for a Civil Engineering Diploma from the Swinburne Technical College. Thus armed I was able to transfer to my home ground in the high country of north eastern Victoria and work on the construction of the Kiewa Hydro-Electric Scheme. The post war shortage of engineers provided a wonderful opportunity to quickly gain a wide range of valuable experience.

Working near the white water streams, forested mountains, and snow covered high country was in complete contrast to the time spent at Woomera. It was varied experience on dam, raceline, tunnel, roads and township construction. In 18 months still as a "Cadet Engineer" I was in charge of 80 men spending the winter on construction above the snowline at Rocky Valley on the Bogong High Plains. Soon after when the shortage of engineers was more acute, I was replacing two senior engineers and in charge of the construction of a concrete dam and also a raceline.

Six years later with a wife and two children I entered the field of Municipal Engineering as an assistant to the City Engineer at Wangaratta. A year later the Council at Corryong in the far North East of Victoria appointed me as their Shire Engineer. It was a golden opportunity as after years of neglect they had a huge backlog of works to catch up on.

On my arrival the modest Shire office was occupied by the Shire Secretary and the Rate Collector, both World War 1 veterans, and both with a wooden leg. Initially the urgency was to get underway with a major road and bridge building programme. The time spent at Woomera was not wasted, as until an assistant was appointed to help I had to do all the surveying and setting out for all new works. Later on I also became involved in the local water supplies, sewerage scheme and town planning. To keep busy the Engineer was also expected to gain additional qualifications as a Building Surveyor and Inspector. There were other minor tasks such as looking after the rubbish collection contract, tree planting, and river improvement works. Being a remote locality without easy access to other professional help it was also appreciated if you could help out local organisations with their projects. That could entail anything from the design of a kindergarten building, a community hall or sporting facilities, to advice on upgrading a racecourse. (All in a honorary capacity).

The shire bordered the Murray River and was on the supply route for the Snowy Mountains Authority works on the western side of the mountains. As a result the township of Corryong underwent rapid growth with the need for major street and drainage works. This also led to the need to locate a site for, and to design and construct a licensed airstrip which was done in record time. We were also deputised to design and build the 'Bringenbrong Bridge' over the Murray River for the heavy transformer loadings going to the scheme's works.

Although not realising it at the time this vast range of responsibility and experience was the basis for later obtaining a position with Western Mining Corporation (WMC) which ultimately would lead to a return to Roxby.

Chapter 8

NEW TOWNS AND NEW MINES - KAMBALDA

The move back to the arid inland came in mid 1968. Western Mining advertised for someone with very broad experience to manage the development of the new township of Kambalda in the Western Australian goldfields. I was appointed to the position with the title of Township Development Superintendent and entered the new world of the mining industry.

The continuing discovery of increasing quantities of nickel ore in the area led to the policy to quickly become established as a reliable supplier of the metal. This meant a rapid build up of the on site workforce who were to be accommodated close to the mining areas. A start had been made with a number of houses, single quarters and a construction camp. These were on the site of the long deserted 1890's "gold rush" township of Kambalda. The land available there for more housing was limited by terrain restraints and the discovery of more ore alongside. This made the development of the completely new town of "Kambalda West" necessary. This was to be about six miles away on a site already selected. Town planning and engineering consultants had made a start on the layout and the design of roads and services. While this was proceeding there was the need to fill up of all vacant spaces in the original town with houses, and to add to the single quarters and construction camp. A large caravan park was also a priority to help provide some of the needed accommodation more quickly.

It was my role to ensure that all of this happened with great urgency and with some very specific ground rules. In particular there was to be extreme emphasis on the protection of the existing trees and vegetation throughout the whole area. In the townships clearing was to be restricted to a minimum both to retain the significant stands of eucalypts and to control dust.

Instead of working under an undemanding council and in sole charge of my work I now found myself as part of a team of mining industry specialists. These were the chiefs of the various operating and construction departments. There was the mining engineer, geologist, metallurgist, mechanical, electrical and construction engineers, accountant and personnel superintendent. Autocratically heading the whole operating and construction effort was the "Resident Manager" John Brian Oliver, (referred to by all as 'J.B.O.') who had first interviewed me for my position. I got to know each of this strange band very quickly as they were all pushing for more accommodation so they could increase their workforce. WMC had not had cause to employ a civil engineer nor to carry out municipal type works of this magnitude previously so it was very much a time of judgement.

It was first necessary to correct the ways of most of the contractors who had been working in the original town now known as Kambalda East. They were erecting houses at a great rate with inadequate foundations, and had copper pipe for temporary water connections criss crossing the area in all directions. The blasting of the shallow rock in the trenches for water mains and to install septic tanks was carried out with very generous amounts of high explosive at all times of the day. This resulted in rocks of all sizes peppering the surrounding populated area. The task of correcting this practice was helped enormously when airborne material broke the tiles on the roof of the Manager's house.

Order was eventually restored after some early difficulties and additional staff were appointed to ensure works were carried out correctly and in an orderly manner. This was the start of becoming accepted in this new and very different working and social environment. It was strange to find a lingering class structure so different from the homogeneous community of rural Victoria. No doubt this was a carry over from the earliest mining where the workforce was structured according to that which applied in 'the old country'. The Mine Manager and his 'Senior Staff' were provided with large houses built on the high ground in groups. Their gardens were maintained by the company and their rentals were nominal. The ordinary 'staff' were housed in smaller dwellings or in catered 'staff single quarters' at a subsidised rental. The married 'award workers' at the time had a long wait before being eligible for housing which generally was to the same standard as for the staff but with lesser fittings. If single they were accommodated in a separate area of the single quarters. All this combined with houses of two, three, four and sometimes five bedrooms made programming and allocation to the most deserving a nightmare. To top it off demand far exceeded supply with each 'Head of Department' desperate to attract essential tradesmen who would only accept employment if housing was provided. The contractor workforce building the town services and houses together with those building the mining facilities, metallurgical plant, workshops and services were accommodated in a separate 'Construction Camp.' All in all there was a desperate need to expand each type of accommodation and to add to the town community facilities. The high labour turnover also added to maintenance needs as units were vacated in a damaged condition.

To assist in relieving the accommodation shortage we quickly designed and organised the construction of a large caravan park. Contracts were then let for the building of the first roads and services in the new township to be called 'Kambalda West.'

The new town was planned with space to accommodate up to and beyond a population of 6000. To create an atmosphere of permanence, a masonry block manufacturer was attracted to set up in the town. This would permit houses, shops, the hotel and other commercial buildings to be constructed with concrete blocks rather than timber frames as used in the original town now known as "Kambalda East".

It was now necessary to come up with a range of house plans so we could call tenders for building the first couple of hundred as quickly as possible. JBO liked to be deeply involved in ensuring that there was no significant variation in the content of each size of house. He was fully occupied during daylight hours so many nights and early mornings were spent coming up with the floor plans for tendering.

The scope of the nickel operations was being increased continually as drilling outlined more and more ore. This built up the pressure as every change meant a new budget had to be compiled for the years ahead. That entailed the gathering of the estimated workforce requirements from each of the departments and budgeting to build quarters to house them. The large bucket of money that the company had started with to get nickel production up and going in a hurry must now have been nearly empty. We seemed to be re-budgeting continuously for several months and trying to pare costs downward wherever possible.

In addition to the townships I was now involved with all of the civil engineering in the area and some of the buildings associated with the mines and plant. There were now three senior engineers on my staff with back up drafting and secretarial assistance. Eric Bingley, and John Dumesney both had been municipal engineers; Eric in Western Australia and John in Victoria. The third, Bert Barnes had been at Kambalda from the start working with a firm of Perth based consultants. Eric was given the task of looking after all of the works at the new town of Kambalda West, and John the old town and the area roads and civil works. Bert's speciality was water supply, tailings storages and later on sewerage treatment and water harvesting and conservation. That left me with the thankless task of rationing out the available accommodation for contractors. This always seemed to be insufficient, and they were each desperate to add more workers to speed their works.

I also seemed to be getting involved in liasing with the various government departments in Perth that were taking an interest in our doings. There was an increasing number of private developers wishing to establish themselves in the new town. This meant a growing amount of time spent with the company lawyers negotiating conditions and drafting agreements. These included a comprehensive hotel-motel, shopping centres as well as service stations and numerous other facilities.

Water was a very precious commodity in this semi-arid area being piped in over a distance of about 550 km from the Mundaring reservoir near Perth. It was therefore limited in quantity as well as being costly. Bert Barnes conceived and engineered various schemes to maximize the benefits of the supply agreement, including storages with low evaporation losses, and harvesting the large quantities of stormwater runoff from the townships. He also used cast off metallurgical plant to treat the town raw sewage flows with over 95% water recovery for reuse. His other speciality was the design and construction of tailings storages and water recovery from these.

In addition to water conservation, steps were taken to rigorously ensure the preservation and enhancement of the native vegetation over the whole area. Removal of a tree without approval became an instant sacking offense. Access to all areas was by defined routes and the myriad of dusty bush tracks were closed and ripped to promote regrowth. Surveyors plotted all trees in the residential areas at Kambalda West and houses were fitted on to plans and allotment boundaries fixed to preserve as many as possible. We received advice (and considerable obstruction) in this regard from a visiting consultant. It is to the credit of Eric Bingly, his supervisors and contractors that so much tree cover and shade was retained.

The company was empowered to develop the nickel project including the townships under a government agreement which was permissive of very rapid progress. We managed to get works under way with a minimum of bureaucratic interference. The Coolgardie Shire Council, with their first major development since the goldrush days were also very supportive. I was even appointed as their honorary Shire Engineer for a short while to help them out with a problem. It was only a few short years when the completion of the 1,000th house in the townships was celebrated.

The provision of the recreation and sporting facilities was also a Company responsibility. When complete these were handed over to the Council free of charge for them to manage. The schools were built by the State and they purchased houses from the company to accommodate the teachers.

Kambalda was receiving world wide publicity so it was inevitable that there were many visitors. These ranged from Princess Margaret and entourage to the Prime Minister and numerous other notables. The inevitable receptions and dinners for the multitude of visitors added to an already busy neighborhood social life. Somehow time was also found to take advantage of the 200 square mile salt lake on the edge of town which had a hard salt crust. A thriving inland yacht club was formed to race fragile, home made, three wheel vehicles at great speed on its smooth surface.

The years at Kambalda passed quickly with an increasing number of new mines and expanded plant requiring roads and services we were kept busy at all times. Frequent visits to Perth became necessary for various meetings including liaison with the Rail Authority on a new line from Kalgoorlie to Esperance. This had a siding at the smelter and a branch line into a load out at the Kambalda plant. We also became involved in providing company housing at Kalgoorlie for the staff at the new nickel smelter.

John Oliver was transferred to Brisbane in 1972 to manage feasibility studies for developing a coal deposit inland from Mackay in north Queensland. Keith Parry his assistant became Resident Manager at Kambalda only to be relocated to Perth a year later as General Manager for Western Australia. At the time WMC was still at the stage where many of the senior staff from around the country knew each other. Living in a company town you associated socially as well as at work.

The work at Kambalda was slackening and at John Oliver's request we compiled a report on a proposed town for inclusion in the study he was managing on the "Hail Creek" coal deposit. This meant going to the site, selecting a location, and spending time in their Brisbane office gathering all the information needed. John had recruited most of his senior staff from Kambalda so it was easy to fit in and a pleasant change of scenery.

Chapter 9

MY POSEIDON ADVENTURE
WINDARRA and LAVERTON

It was on a short visit to Perth for a conference in April 1973 that I was stopped in the street by Keith Parry who happened to be driving past. After arranging to have dinner with us that night he instructed me to attend a meeting at the WMC Perth office early the next morning.

I never returned to work at Kambalda.

Next morning bleary eyed from the night's excesses I learned that WMC had taken a 50% interest in the Windarra nickel deposit with Poseidon Ltd. The meeting was with the Architect they had engaged and Buzz Myers, Poseidon's Managing Director. We were viewing for the first time the Town Plans as prepared by the Architect. I was expected to give intelligent comment on these in the presence of the the high powered attendance. The plans were not exactly to my taste. However I must have passed reasonable comments for I was immediately seconded or loaned to Poseidon with the high sounding title of "Town Development Manager." The next day I was working from Poseidon's temporary office in Perth and arranging to take up residence in that city. The following week was the start of two years of commuting to the mine and town on the edge of the Great Victoria Desert for two days each week in a light aircraft.

The task covered the redevelopment of, and provision of housing in the near deserted old gold mining township of Laverton, 800 km to the north east of Perth. The re-subdivision and servicing was carried out in conjunction with the State and the Laverton Shire Council. It was fraught with difficulties. There were endless meetings dealing with the numerous State authorities, and endeavouring to find answers for the welfare and housing needs for the nomadic aboriginal population. The isolation and lack of facilities added to the problems of attracting contractors and a workforce to the area. To top it off raging inflation was increasing costs dramatically and we had an tight and inflexible pre-inflation budget.

One bright spot on my first visit, was when I entered the one room Post Office. On the wall was a framed cartoon illustration drawn by Len Beadell who had apparently been a regular caller during his road building years.

I was helped along tremendously when Max Hannel, a young architect from Perth was appointed to my staff. The pricing restraints gradually limited the sizes and scope of the houses and we turned to transportables and prefabricated construction.

There were some memorable moments. At budget time on my second visit I had flown to site at 6 am ex Perth, then spent an unbroken day and night with the Chief Accountant pouring over figures. He finally delivered me close to my allocated room in the single quarters at the mine at 2 am. On opening the external door and turning on the light it became evident the bed was occupied and the occupant was in no mood to be disturbed. I was left standing in the freezing darkened camp with my overnight bag not knowing where to find the camp foreman to get another room. The only recourse was to stumble through the saltbush in the starlight trying to avoid the rabbit warrens to the dozen or so temporary houses a hundred or two metres distant. They were also in darkness except for one dim light which went out as I approached. Unaware of who lived where, I selected this house as my target and attended by unfriendly barking dogs found it to be the accountant's. Clad in his dressing gown he returned me to the camp where the camp foreman eventually located another room.

Another was when a pall of smoke in the near distance was the sign of a scrub fire in the lush growth from several good seasons. Bob Floyd, the Resident Manager, decided to check out the danger to the mine area with an aerial inspection in the charter aircraft waiting to return us to Perth. Flying very low they flew along the line of fire which was burning extremely fiercely in dense spinifex. The flames and heat were rising high above the ground and must have vaporised the fuel supplying the the aircraft engines which stopped without warning. It was either come down in the path of the flames or on the hot ashes behind them. Electing the latter the pilot headed in that direction and was much relieved when just on touchdown the engines started up and they could gain height again.

Bob was continually pressuring us to get his 'Manager's House' in the town finished. We had difficulty finding a contractor for the job until we got a price from one of the firms who had been working at Kambalda for years. It was to be constructed with prefabricated polyurethene wall panels and delays in delivery caused the contractor to fall behind. Eventually he had all the panels on site and standing in position ready to tie them together with the roof trusses. At that moment the ill fortune that had been dogging us struck again in the form of a very violent " Willy - Willy." It passed through the town missing everything except the vulnerable, under construction manager's house. The newly positioned wall panels were flattened and damaged beyond repair. The manager wasn't able to move in for Christmas.

In late August of the following year I took time off and with two of our young sons we joined a party driving from Perth to Laverton and into the bush to Warburton Mission. We then continued along Len Beadell's roads to Giles weather station, to Ayers Rock as it was called at the time and on to Alice Springs. The return journey was south to Coober Pedy where it rained heavily overnight. Without 4WD, we barely made it to Kingoonya in a very muddy state by dark the next day.

The Kingoonya Pub was as I remembered it with the addition of a couple of motel units which we quickly booked to avoid setting up camp. The hotel was operated by relatives of the Crosbys and I learnt that Harold Watts had married daughter Doreen and they lived in Melbourne.. Also that Ossie Osborne was managing the North Well Sheep Station a few miles out of town. A phone call and ten minutes later there was Ossie. That put an end to an early night as we relived the past into the early hours.

Back at Laverton the climax of all the activity came when the 'New' Town was officially opened by the Premier, Sir Charles Court with all day activities. There was the bush race meeting in a raging dust storm and the official opening at night climaxing an 18 hour day. The only person not totally exhausted was the Premier who had been on his feet all day.

Chapter 10

BACK TO WMC AND THEN THE MOUNTAINS

I returned from being on loan to Poseidon to work from the WMC office in Perth with the high sounding title of 'Group Civil Engineer.' This time I was attached to the WMC Engineering Services group, (WES), headed by Peter Webster. The WES Engineers looked after the specialised needs of the scattered mining and processing operations and assisted in feasibility studies on new projects. I was kept busy with work at most W.A. operations, on a wide variety of tasks ranging from subdivisions and housing, to exploration for groundwater supplies and tailings storages. I even spent a fortnight on Christmas Island in the Indian Ocean to advise on upgrading the accommodation at the phosphate operation. Then there was the initial development of water supplies and housing for the ill fated Jurien Bay mineral sands operation. This was the only WMC mine situated near the coastline far from the arid inland but was closed after a year or two of production difficulties. Following further studies WMC also relinquished it's interest in the Hail Creek coal deposit in Queensland.

One of the more intense investigations centred on the infrastructure proposals for developing the Yeelirrie Uranium deposit situated about 500 km north of Kalgoorlie. After the areas had been outlined for the mining, plant and tailings, and the workforce needs estimated I had to select the location and layout for a town, and access routes. Another task was co-ordinating a water exploration programme for the project which was being undertaken by Cecil Forbes of Australian Groundwater Consultants. All of this was needed for feasibility studies and so that consultants could prepare an environmental impact report for publication and exhibition. Although eventually passing all tests including detailed metallurgical testing, the project stalled because of Government restrictions on opening up further uranium mines.

I was particularly interested when in 1975 WMC announced the discovery of copper in a deep drill hole near Andamooka in South Australia. This was in the region we had mapped when I was in the army. In Perth office we eagerly waited the long intervals between the announcements of subsequent drilling. The succession of barren holes and persistence was finally rewarded with the significant results from the tenth drill hole.

Shortly after we were to compile a preliminary study of the works and costs of developing a hypothetical mine at the site of the discovery and to predict what works and infrastructure would be needed. It gave me the opportunity to visit Adelaide to seek information on services and available maps. I also caught up with Len Beadell for the first time since 1947. A month or so later I made a day trip from Adelaide to the drilling site now known as being on the Roxby Downs Station and adjacent to one of their watering points called Olympic Dam. Landing at Woomera gave me the first opportunity of seeing the transformation of the dirt airstrip we had left in 1947. It was also my first look at the town, thirty years after we first surveyed the barren townsite.

Travelling with John Emmerson, the drilling supervisor we followed the station track to Roxby Downs station homestead, now a new building on high ground. It was then on along more narrow dusty tracks,via a remote drill site, to the drillers caravans at Olympic Dam. A quick tour of the drill sites and it was time to return to Woomera and the city. The visit refreshed my knowledge of the terrain and hopefully equipped me to better assess what works might be required to support any future mining operation. The assumptions then made were roughly costed and incorporated into the very approximate preliminary feasibility study. Of one thing I was certain - if I had anything to do with selecting a townsite it would be in an area with some trees, bushes and soil and not on bare barren gibber like Woomera.

In late 1978 with the downturn in metal prices and virtually no work to do, I resigned from WMC and we returned to Victoria to settle on the family farm. Back in the familiar mountain country we lived in a caravan while our house was being built. Life was peaceful and idyllic but was rudely interrupted early in the new year when I was asked to advise on housing a mining workforce in Tasmania. This involved an enjoyable week on the island state and meetings in Melbourne. This was followed with a few days trying to locate a route for access to a proposed mine in the Victorian Alps for WMC. Then a return to Perth to assist on another study. It was well into the winter when things quietened down, the walnuts we grew on the farm had been harvested and we could move into the new warm house. Consideration could even be given as to how the day might be occupied.

I had a chainsaw in hand when I received the message to call John Oliver now back in the Perth Office. I rang him.

"Would you be able to go to Adelaide for three months to establish a camp, airstrip and water supply at Olympic Dam?"

It sounded reasonably straight forward after I spoke to George White, the Geologist in charge of the exploration, when I rang him at his Adelaide office. I had known George when we were both at Kambalda and saw no problem in working to his requirements. He explained that Olympic Dam Exploration was to become a Joint Venture, The incoming partner, British Petroleum, was to meet the costs of a greatly increased drilling effort. This needed the establishment on site of a fully equipped camp for 60 men, and the provision of water supplies and an airstrip, all before the New year in four months time.

On September 10th 1979 complete with family dog we moved to Adelaide for three or four months.

It was more than twelve long years later, on final retirement that I returned to the 'the house on the farm'.

BOOK THREE
1979 to 1991
RETURN TO ROXBY DOWNS

Chapter 11

GETTING STARTED

In September 1979, the office for the WMC staff engaged on the Stuart Shelf exploration, of which the Olympic Dam discovery was a part, was extremely modest. It consisted of a smallish house on a main road in the Adelaide suburb of Marion. The association of uranium with the ore discovery was obviously sufficient cause for the local "greenies" to daub the house fence with their anti messages for all to see. All rooms in the house were fully occupied by the several geologists and draftsmen connected with the exploration. My only option for an office was to occupy some unused space in the garage. The one redeeming feature was that the Marion Hotel was just across the road.

On arrival in Adelaide I settled into the garage and began making initial contact with various government departments and potential suppliers of camp buildings. I then had a visit from John Oliver and Bruce Gardiner from Perth office. Bruce was the Administration Manager with a reputation for dealing with the near impossible and making on the spot decisions.I learned that the future works connected with the Olympic Dam discovery would be managed by a new entity called "Roxby Management Services." (RMS). This would be a Western Mining company which would commence operating in the New Year.In the meantime I would be working on a contract basis for the Exploration Division of WMC. My main task being to get a 60 room camp, complete with services and an airstrip ready for use at Olympic Dam by the end of the year. This was to enable a greatly accelerated drilling effort to get under way on site in the first weeks of January. I was also given another small task. There was an immediate need for a larger office to relocate the staff from Marion to a near city location and to provide for the increased number that would soon be needed. This was to be a temporary move until a permanent home could be obtained for all the WMC groups scattered around the Adelaide suburbs. The same day we inspected a well situated, newly completed building at 205 Greenhill Road on the city fringe. We regarded it as a suitable interim office for Roxby Management Services and leased it on the spot. I was issued with an official order book to get started with the fitting out of the wide open space in this new building. I immediately took advantage by moving in and managed to get a phone installed. It was a mixed blessing as the number given to me had obviously been recently held by another subscriber. Until we had an office exchange in operation some weeks later at least half of the callers asked if that was "The Home for Incurables."

Bruce was back a week later when we met with the manager of a large Adelaide building firm who had plans to build offices on a vacant block at 168 Greenhill Road. We looked at their plans and it was again agreed on the spot that we lease at least one and a half of the two large floors on completion.

Once again I became involved being detailed to liaise with the builders as the building progressed and to negotiate any changes or additions that might be required.

The last few weeks in September were busy taken with feeling my way into these various tasks and getting to know the many people and organisations with whom I would be dealing. Fortunately I knew many of the staff at Kinhill Engineers, locally based consultants, who I had worked closely with on the Poseidon project at Windarra. They expressed a willingness to help in any capacity and to provide an "on site" supervisor when required to look after the installation of the camp and services. The Joint Venture's intensive drilling programme at Olympic Dam, timed to start immediately after the new year, would increase the number of drill rigs on site from two to double figures. I lost no time in calling for prices for supply and installation of the 60 room camp. This to be complete with water, power and sewer reticulation all on a hypothetical location to the west of Andamooka. The camp would be required to be self contained with a Kitchen-diner building with freezer, cool rooms and large pantry storage. A separate unit would contain a recreation room, wet canteen with bar plus cool room, and an office and first aid room. The bunkhouses were to be to of standard design, but upgraded with full insulation, electric heating, air coolers and termite resistant framing. Toilet, laundry and washroom facilities would be in separate hard wearing buildings. I drew up a camp layout plan that could be adapted to most sites as a basis for pricing.

I then visited Detroit Engineering, regular suppliers to WMC, and asked them to urgently prepare prices for the supply, installation and commissioning of a diesel engined power plant. This was to be complete with control room and equipment for automatic operation, and installed on site by early December.

It would be necessary to ensure that the campsite was suited to easy expansion and reasonably close to the airstrip. It had been decreed that the airstrip should be capable of being developed for aircraft up to DC3 standards. There was also the need to have a large site close to the camp for the handling and storage of all past and future drill core. All that then remained was to go to Olympic Dam and select a site for our new village.

Chapter 12

A SPOT FOR A VILLAGE

George White and I set off to drive to Andamooka on October 1st armed with the small scale and outdated maps and aerial photographs of the area to the north of Woomera. Most of these dated from our army mapping of 1947. I also had my newly issued security clearance permitting me to enter the "Woomera Restricted Area." The eastern boundary of this area was on the fence line separating the pastoral leases of Andamooka and Roxby Downs. It was a significant line with the whole of Roxby Downs Station and the area for the proposed intensive drilling being to the west and therefore inside the Restricted Area. We were intent on keeping the camp and airstrip out of this area to avoid potential delays in obtaining approvals from the remote bureaucracy in Canberra. This meant locating the developments to the east of the fence line on the Andamooka pastoral lease.

Travelling north, I was pleased to see that the road was sealed for most of the way from Port Augusta to Woomera, a significant change from the two dusty wheel tracks of 1947. The way through Woomera was blocked with security gates and a narrow dirt track detoured around the perimeter fencing to my old stamping ground at Phillip Ponds. The fine old stone coach-house dating from the 1880's had been senselessly vandalised almost to the point of destruction. What had been an excellent example of the ability and tenacity of the early pioneers was now a roofless gutted ruin.

The 'road' from the Ponds to Andamooka was if anything worse than the dirt track of 1947. Following outback tradition travellers had bypassed the many mudholes and hollows on the original route by swinging around them on to the solid edges. The new bypasses then became boggy and the road became wider and wider. There had apparently been recent heavy rain in this driest part of this driest State. George who was driving the already uncomfortable 4WD vehicle had to quickly choose whether to go around on the rutted bypasses or bump straight through the succession of ponds. It was quite dark when about 15 miles from Andamooka we came across a muddy sedan hopelessly bogged in the middle of a larger than normal puddle. The driver, a bank manager on his routine visit to do business at the Opal Fields was quite pleased to see us. He waded into the mire, collected his suitcase, and travelled on to the Andamooka Motel with us. His vehicle was left all alone to be retrieved next day.

On arrival at Andamooka we were warmly welcomed at the then one star motel, amply fed and enjoyed a well earned drink with the local patrons of the crowded bar.

We were up at daylight and headed off towards Olympic Dam after a tinned breakfast. Andamooka had progressed since my previous visit in 1947 with a number of above ground buildings now evident and many more mine shafts dotting the landscape.

We followed the two wheel tracks heading west firstly over the open gibber extending out from the opal fields, then for some distance along the edge of a single sand dune. Eventually crossing the dune the tracks led us down a lengthy open slope to a fenced area enclosing a large excavated pastoral dam locally known as 'Twelve Mile' dam. It was filled to the brim with clear water collected from runoff from the surrounding catchment.

Continuing on we entered a wide swale with east-west trending sand dunes covered in scrub slowly closing in on each side. Shortly after we came to a fenceline with an unlocked gate festooned with signs advising that we were on the boundary of the Woomera Prohibited Area. Disregarding the fine print with the legal jargon, we entered and drove on with the track now twisting its way across mulga flats between closely spaced dunes.

Thirty kilometres from Andamooka we reached the cluster of three or four battered caravans that housed the drillers working on the site. These were parked among some sparse mulga at the edge of a small dry claypan. This being where the now famous pastoral dam had been excavated at the time of the 1956 Olympic Games in Melbourne. All rural watering holes having to have a name, the pastoralist with true bush logic had therefore called it 'OLYMPIC DAM'.

The dam was full, with extremely muddy water collected from the claypan and surrounds. A stumpy, aged windmill was sited to pump this muddy brew to an adjacent tank which fed on to a long open trough. The cattle grazing the surrounding country had access to this for their daily drink but were fenced off from the dam to prevent them trampling the soft edges. The drillers did not depend on this water for the camp. They carted in their modest needs on a weekly basis from the large clear water dams at the Roxby Downs homestead supplied courtesy of the owners.

The drillers also had a bore equipped with a diesel driven pump tapping the highly saline aquifer underlying the area at a depth of 50 to 200 metres. This was tankered to the drill sites and used to mix the "mud" circulated in the drill hole to float the drill cuttings to the surface.

We made a quick inspection of the operating drills and the general area to the north which would be the focus of the intensive exploration starting in the new year. The site was crossed with stable rich red dunes generally trending east-west and separated by swales of varying width. The saltbush and mulga cover on the swales changed to shrubs of various types on the dunes. Access from the camp was on a narrow dirt track coated with local calcrete gravel at the dune crossings. Cross tracks in the swales led to the well spaced drill sites.

After this brief visit and paying our respects to the camp occupants we retraced our steps back through the security gate on to the Andamooka Station land. We spent the rest of the daylight hours bouncing across gibber and dunes seeking a spot for the village roughly following our course with the aerial photos. We traversed along open areas that might be suitable for an airstrip reading the vehicle speedo to check the clear lengths. To the north, the dunes became higher and closer together so we concentrated on the more open country to the south and the east.

To avoid excess travelling from the future camp to the mineralised area for drilling and possible future mining we wanted to be in the range of 3 to 5 km from Olympic Dam.

We covered a lot of country but there was not a great deal on offer. A small confined gibber covered area of inside a horseshoe of low dunes and with a few scrawny mulgas seemed to be the best available. We marked the location on the aerial photos together with an airstrip site on a larger stretch of gibber some distance away. Overnighting at Andamooka, we returned next morning for another quick look and to double check before driving the 600 km back to Adelaide.

Chapter 13

THE OUTBACK ROAD BUILDER and WATER

It was apparent that an access road from Andamooka to Olympic Dam for the transport of the prefabricated camp buildings and heavy equipment to site was of high priority. We would have to be content with the existing 'dry weather only' road from Woomera to Andamooka.

Back in the office we made contact with John Emmerson the WMC Drilling Supervisor based at Woomera. John was still looking after a number of drills busy on exploration on the extensive province to the north of Port Augusta of which the Olympic Dam area was a small part. In accessing and preparing drill sites John would have had to make use of earthmoving plant. His advice and offer of assistance was to prove invaluable. He agreed to contact "Northern Earthmovers," a three man outfit with a range of plant and headed by one Mick Collins to check if they could help. Their speciality was dam sinking anywhere in this region of the outback and they were well known to John as reliable, competent and honest.

A few days later, under John's guidance and direction as to the appropriate route and standard of construction, they started building a solid outback road in a westerly direction from Andamooka. Thus started a working relationship that was to last many years. I caught up with this self contained little group a week or so later and found Mick to be a solid man of the bush, rugged and weatherbeaten. He had served with an airstrip construction group during the war and had his home and base at Iron Knob, a short distance from Port Augusta. Mick was ably assisted by the other two plant operators in his party - his son Ron, a big lad, well over six feet and of ample proportions and Danny, quiet with a terrific sense of humour and a perpetual wide smile. Danny was of Aboriginal descent, originally from the area to the east of Lake Torrens.

In the following weeks there was never a dull moment. The tenders for supply and installation of the camp buildings and services were received and the contract let to 'Atco', a firm based north of Adelaide at Elizabeth. The contract for the power plant was also let, with provision for soundproofing of the diesel-electric sets as they, of necessity, had to be close to the camp.

Water supply was the big worry. Tom Allison from Roxby Downs Station who was supplying the drillers agreed to let us cart a limited amount from his house dams to cover needs of the workforce installing the camp. We ordered a couple of prefabricated 2000 gallon concrete tanks to go to site as soon as the road to the camp area was complete. That left the problem of the larger long term needs after the camp was occupied.

The "12 mile dam" on Andamooka Station which we had looked at on our visit looked very tempting as only a very small amount of the water in storage would be used to supply the needs of the stock on the property. It was full from recent rains but without further falls by far the greatest proportion would be lost to evaporation over the following two to three years. We reasoned that if we could save or reduce this loss we might have a good case to present to the owners to access part of this. The only other option, and necessary for the longer term, was to seek the approval of the Commonwealth authorities to draw from the Woomera supply. This was at the end of their 200 km ageing pipeline from Port Augusta which in turn was supplied from an even longer State pipeline from the Murray River at Morgan. Obviously the cost of pumping this water to Woomera was very high, and the addition of transport on to Olympic Dam would increase costs considerably.

I asked Kinhill Consultants if they could come up with a scheme to reduce the evaporation from the station dam to ensure the continuing requirements for the cattle and the immediate needs of the camp. The Station owners agreed to discuss any proposal we might wish to present.

We also advised the Mines Department and The Pastoral Board of the sites we had selected for the camp and airstrip seeking their blessing for us to proceed. They were sensitive to the location on Andamooka Station as the property was already adversely affected by the extensive area of their land taken up by the opal mining. We in turn were were emphatic that we wished to be clear of the restrictions of the Woomera Restricted Area. The Department of Civil Aviation were also contacted regarding the airstrip. They saw no problem as long as aircraft kept clear of the Woomera restricted air space at certain times.

Chapter 14

A NEW SITE FOR THE VILLAGE

On October 10th Rod Everett from the Pastoral Board contacted us with the news that the Commonwealth had agreed to change the Restricted Area boundary. This would enable us to locate the camp and airstrip on Roxby Downs and would also free up the mineralised area where the intensive drilling was to take place. Apparently Rod had pursued the matter at a high level resulting in the State Premier contacting the Prime Minister and receiving almost immediate approval. This was a very timely and significant decision which enabled the siting of the initial and subsequent major infrastructure in the most favourable locations.

There was then a very swift change of our plans and the speedy organization of a charter aircraft to fly to site the next day. On board were Bob Perry from Kinhill, George and I. My task was to select new sites for the camp, airstrip and core farm all in the one day. George would be fully occupied with geological matters and Bob was to check out the quantities of water in the various pastoral dams. We had previously organised for Peter Simmonds, an Adelaide based surveyor to travel by road to start preparing contour plans for the original campsite and airstrip locations. He with his chainman would conveniently have arrived on site and would be camping in the nearby bush. They could now help Bob measure and depth the water in the dams with the aid of a small blow-up rubber boat we had brought with us. The drillers were advised of our visit via the WMC Exploration Division radio system. This enabled us to have Sam the camp foreman meet us when the aircraft landed on the dry surface of a large claypan, Lake Blanche, about 12 km to the west of Olympic Dam.

We left Adelaide at daylight in an 'Emu Airlines' aircraft and after dodging around the restricted airspace at Woomera finally made it to Lake Blanche. Sam had sensibly marked out a runway with some strategically placed camp chairs, He had carefully picked the line where the surface was least cut up with cattle tracks leading to the large dam on one side of the claypan. We crowded into Sam's vehicle and headed along the dusty station track to Olympic Dam, where we split up to go our separate ways. Armed again with aerial photos and a 4 wheel drive vehicle I headed off to find a new campsite. The only sensible option was to check out sites to the south. By keeping reasonably close to the station boundary we would cause least disruption to the pastoral activities. It was also in the direction of Woomera where any future direct road to a mine development would terminate. By this time the incoming road construction from Andamooka was nearing the station boundary fence and the first camp buildings were due in two to three weeks.

After crossing six or seven low dunes the landscape started to look promising. There was a long straight swale between spaced east-west sand dunes that looked as though it could possibly be long enough for the airstrip. At the eastern end the swale opened up to the south into a large amphitheatre of gently rising ground. This contained a significant stand of large myall trees backed by a wide elevated mulga covered sand plain. I lost no time checking this area out on foot and determined that the drainage slopes and soil conditions were ideal for the camp. The soil was a competent red sandy loam and trenching was very unlikely to encounter rock.

The next move was to run the vehicle along the line of the possible airstrip with speedo readings to see if the clear length was adequate. It was rough travelling and as a consequence slow. The open area of the swale was covered with saltbush clumps and numerous dead mulgas underlain with logs and debris. The length required for light twin engined aircraft came up on the speedo on level grades and with side clearances looking good. I continued on up a gentle slope to determine that additional length could be developed with a minimum of earthworks. A potential gravel deposit adjacent to the western end of the strip also was of interest. That done, the likely sites were marked on the aerial photos and I proceeded further south to see what lay over the next dunes. It was again promising with a succession of open swales and low dunes, all with reasonable tree cover.

Returning to the drillers camp well satisfied I briefed the surveyor to establish survey control points in accordance with Australian mapping and level datum standards at the airstrip and camp sites. Then to take levels over both areas for the preparation of contour plans. A message was also left for John Emmerson. This was to build a road from Olympic Dam due south to the new camp site and to extend the road from Andamooka to tee into this.

The other plane passengers had also completed their tasks including measuring the quantities,and depths of water in the various dams. It was then back to the aircraft at Lake Blanche and the return flight to Adelaide.

The next day in company with John Collins, one of George's geological draftsman, a visit was made to the Lands Department mapping section. John was well known to the staff who had always been helpful in the past when we ordered aerial photographs. Our immediate request was to see if the area of the proposed camp and airstrip on the very small scale photos could be enlarged enough to be of use in planning the camp layout. We also wished to check if some new larger scale colour photography encompassing Olympic Dam could be provided in the near future. They indicated that they could arrange for the lines of photos to be taken next time their aircraft was passing by on a run to scheduled work in the north of the State. They would also take a high level photo covering the whole area on one frame.

They were a bit hesitant as to whether the postage stamp size area of the campsite on the old black and white photos would enlarge sufficiently to be of use but would try. We stressed that it could be of great help to us in plotting in contours and the position of camp buildings and services. We were intent on saving all the myalls that made that site stand out and would provide much needed shade. Leaving them with an official order for the work we awaited results. They accepted it as a challenge and a day or so later we collected some very coarse grained enlargements. These had been blown up to an approximate scale of 1 to 500, covering the site for the camp and most of the airstrip. They were magnificent. We could see the spread of canopy of each tree, the extent of the bounding dunes and the open spaces suitable for access and buildings.

We ordered more prints of the enlargements, which I took to site and had the surveyor plot contours directly on to them. I received these a day or two later and added sitings for the bunkhouses, washrooms, kitchen-diner, recreation room and canteen as well as the access roads and services. The overall plan of the camp was then drawn up with corrections as necessary to the minor distortions of scale on the photographs. This was followed by the detailed plotting of an overall plan of the airstrip area with contours. These in turn let us start to compile composite plans at a smaller scale covering wider areas and leading to the production of 1:1000 and 1:5000 standard mapping sheets.

The immediate concern then became fixing the centre line and ends of the airstrip and the position for the terminal so we could get construction under way. Another visit to site was called for and we again landed on Lake Blanche to be picked up by Sam. George visited the camp site with me and we checked out that the swale over the high dune to the south was suitable for the Core Farm and a Drill Workshop. The power plant and control room would also nestle into the south foot of the dune which would protect the camp from the noise. I was able to fix the airstrip centre line and ends, and pick a location for sewerage treatment lagoons. These would be at the lowest part of the swale and allow gravity flow in the outfall pipe from the camp.

Although the site drawings were incomplete, these and the layouts on the photo enlargements were sufficient to set out and check the locations for the camp buildings, roads and services. There was the flexibility to make minor changes as necessary.

The airstrip centre line was then checked and pegged at each end. The obvious siting of the airstrip terminal was at the extreme eastern end of the landing strip where it would be within easy walking distance to the camp.

Mick Collins and his team had now finished the road formation into the area and were heading south to the campsite on a line that would skirt the eastern end of the airstrip.

All that was now needed was a competent supervisor to look after the camp site works, building and services installation and airstrip construction. Kinhill kindly obliged by making available to us Brett Allen, a young New Zealander with broad works experience. Brett was ideal for the job and on October 24th with a hired 4wd vehicle drove to Andamooka and took up residence at the motel. By then the new road was nearing the campsite.

I joined Brett on site when he arrived and provided him with the latest drawings and explained that the first five bunkhouses were due to arrive in the first week of November. The first priority would be to complete the road into the camp and prepare the sites and access for the buildings. The road builders were happy to continue on when they finished the work at the camp. He could then start them on clearing the site for the airstrip.In the absence of a reliable means of communication to site it was agreed that Brett would contact me each night from the public telephone at Andamooka.

Two days later George and I returned to Olympic Dam, landing at Whyalla on the way to confer with the owners of Andamooka Station. We presented a case to store the water in the 12 mile dam under a floating cover so as to prevent the massive evaporation losses. The relatively small quantity needed for their cattle would be met from the safely stored supply with the balance sufficient to supply our camp for about eighteen months. We would ensure that a reserve was maintained for the cattle beyond the time when the dams would, without further inflow, have dried out naturally. Such a scheme would have met our immediate requirements at a lesser cost than carting in from Woomera and without detriment to the pastoral needs. However they were not convinced so we had to look at the other options.

Desalination of the salty water in the quartzite underlying the regional area was not feasible due to the high level of sulphates in the salt content and other factors. The only assured source seemed to be tankering in Murray River water from Woomera, or failing this from Port Augusta.

Brett had settled in at the Andamooka Motel and had learned that there was quite a range of construction type plant scattered around the opal mines. This looked promising if we could arrange to hire some to help build the airstrip. The centre line of the runway was at the low point of the hollow between the spaced dunes. The natural surface across the strip was therefore concave and large quantities of fill would be needed to build up along the centreline to establish drainage grades to the sides. Brett agreed to determine the types of plant that was available and the hire rates for each. He would let me know on his nightly call from Andamooka. I in turn was to see if Telecom could establish a phone link to the camp site as a matter of urgency.

The other immediate need was to have the two prefabricated concrete tanks we had ordered delivered as soon as a track into the campsite was trafficable. This would let us establish an initial water supply for the contractors by carting in limited quantities from the Roxby dams. We kept looking at the muddy water in Olympic Dam which was of good quality if only we could find a way of cleaning it up. The colloidal clay making it muddy was too fine for filtering so we tried an old bush method on a small sample. Adding a small quantity of salty water was supposed to settle out the mud in very muddy water. We took samples of salt water from the drillers bore and the muddy water from the dam and tried it.

It worked. After trying various mixtures we soon found the minimum requirement of salt water to drop the silt out and that the clear water obtained was acceptable for use. That gave us the opportunity to meet the immediate needs of the camp while we sought approval to draw from Woomera. This approval was forthcoming shortly after a visit to Woomera and a meeting with The Administrator and some of his staff. To prepare for supply we had to provide a metered pipeline from their mains down to a fenced enclosure with storage tanks and loading facilities at Phillip Ponds. All this would take time so we tried out cleaning up the Olympic Dam water on a larger scale in a hole dug in the claypan alongside the dam. Heartened by the result we calculated that the water in the dam could keep the camp in water until the end of the third month of the New Year. Heartened by this we ordered rolls of polyethelene pipe to feed from the dam to the tanks at the camp.

On November 5th the first bunkhouses were complete and ready for inspection before loading out for transport to site. We had added to the Atco basic specification asking for furnishings and other improvements and were well satisfied to let them be loaded for transport.

Then the inevitable - it rained from Port Augusta to the north. The next report was of the loaded transporters bogged at intervals from just south of Woomera to Andamooka. A day or two later the first units arrived on site to pads already cleared and prepared under Brett's direction and Olympic Dam Village was born.

The Atco Company workmen installing the prefabricated units worked long hours and camped in the empty bunkhouses. They were soon joined by others to install the sewer and water lines and the large squatters tank. Then came the prebuilt sections of the kitchen, cool room, freezer, store and dining room. These were soon followed by the ablution - washroom units, canteen, 1st aid room and camp office. As the settlement started to take shape in the increasing temperatures at the start of summer the three diesel electric generating sets arrived with the control room wired for automatic operation. Other teams were now busy connecting power supply to the new buildings and putting up poles for area lighting.

Telecom had by now provided a phone line with a radio connection through the Andamooka exchange and were preparing to increase this to a six line system. This meant that we needed a small manual exchange and cabling to various buildings and a couple of public phone boxes

Brett was kept busy keeping up with all this activity at the camp but still managed to look after Northern Earthmovers who had finished gravelling the road into the camp and over the dune to the core farm site. He then had them clear the light vegetation from the airstrip area. I was now flying to site on a weekly basis taking with me the latest copies of the plans which were being updated continuously. When the surveyor provided all the levelling data necessary I was able to prepare a drawing showing the detail necessary to construct the airstrip. For most of its length the centre line was dead level and it was only a matter of filling to a depth of a metre or so to slope uniformly out to the sides for perfect drainage. The runoff from the camp would drain towards the south side of the strip, so most of the fill would be obtained from borrow pits dug along this side.

The airstrip construction had started in earnest as soon as I finalised the design and provided Brett with the detailed plans. He had gathered a wide range of plant from Andamooka which was soon moving dirt for us. The opal miner owners had leapt at the chance to earn at an hourly rate. They had somewhere among the multitude of mines found tractors, scrapers, loaders, water tankers and tip trucks. These added to the plant of Northern Earthmovers soon had earth fill being placed at a great rate. It was extremely dusty and a major difficulty was supplying enough water to compact the fill and to maintain visibility. George had organised a drill to the site as soon as it was selected to ensure that there was only barren rock underneath and no rich mineralisation. We were able to use this drill hole to advantage by pumping salty water from it for the construction. This was supplemented by the supply from the drillers bore and by the rain that had bogged the trucks bringing in the camp units.

In mid November Peter Webster, Ted Winship and Bob Crew from the WMC Engineering Services group in Perth arrived on a visit to Adelaide office. They had all contributed to the 1977 preliminary study but had not been to the site. George and I escorted them for their first look at Olympic Dam, the drill sites, the camp and airstrip work. A week or two later John Oliver also came to visit on a day when it was well over 45 degrees. On returning to the aircraft at Lake Blanche where it had been standing in the sun all day we found the inside to be like an oven. The pilot took the plane into the air and much higher than usual so that we were gasping for air, before he was persuaded back to a lower altitude.

There was growing interest in what was going on with trips to site each week usually with other visitors to show around as well as checking on the work in progress. On the occasion of visits with officers of the Premier's Department, and then the Mines Department we landed at one of the constructed airstrips at Andamooka. The Mines Inspector based there was then able to help with transport along the new road and around the site.

We were also busy in Adelaide moving in to the interim Greenhill Road office, and settling in additional staff. We had partitioned off a number of offices and added to the available furniture. As well as offices for George, who was to look after Olympic Dam exclusively, there was one for Geoff Hudson who was to manage the continuing exploration in the rest of the northern region. The two of them had appointed two young ladies for secretarial and office duties. Linda and Jodie also had to come to terms with and manage our newly acquired switchboard with its mass of cabling and plugs. Shortly after, all of the staff from the Marion office had moved in and settled down to prepare for the increasing workload in the New Year.

The need to move the accumulated drill core of past years from its present location down south into our planned core farm was now considered to be urgent. Added to this George was anxious to have a Drill Maintenance Workshop and a comprehensive Sample Preparation facility built adjacent to the core storage yard. The former was to be a solid steel frame building with crane rail which we managed to outline on a large sheet of graph paper. Within a week we had a price from Atco to fabricate and erect the building on site and have it completed within a month. We checked with the WMC Engineering Services group in Perth { WES } that the price was competitive and wrote an order for Atco to proceed.

The Sample Preparation facility was much more complicated and not a very common item. It was arranged that I visit the one that WMC had at Ballarat to find out what was needed. It would be convenient to do this by calling in on the Friday before Christmas when driving back to Victoria for a short break.

The result of all these happenings was that the paper work was starting to build up at a great rate. So did the number of personnel requiring desks in the Greenhill Road office. WES sent over a contracts officer, Alan Davies, to help handle accounts and payments. The contract with Atco for all the works at the camp had been briefly documented and was straight forward. Service lines were priced at a rate per metre to be measured up on completion and paid accordingly. Buildings were priced individually so it was a simple adjustment when during the initial work we were asked to increase the capacity from 60 to 80 rooms. The kitchen, diner and canteen were landed on site in three metre wide modules. We had allowed plenty of room around each building and they eventually grew from the small beginnings to very expansive installations. We kept track of costs being incurred on the road and airstrip works on a daily basis. Brett would advise of the hours each item of plant worked when he rang each night and confirmed this with weekly time sheets. It was a simple matter to get a reasonably accurate and up to date figure on costs incurred when the need arose. There was a large amount of trust about the whole exercise with the expectation that the work would be managed promptly and economically. The expanding staff numbers in the Adelaide office soon included accountants and the paperwork grew exponentially.

The work on the airstrip continued apace with the earthworks rising towards final levels. The low ridge at the western end yielded some excellent gravel which was used to surface the runway and terminal area as well as the roads in the village and to the north.

December progressed, the power plant was installed and commissioned, we now had a six line radio phone connection, and a mini switchboard. There was water in the small concrete tanks, and the squatters tank was complete ready for filling. Two small sewerage treatment lagoons had been excavated on the far side of the airstrip and the outfall pipeline from the camp installed.

A very necessary approval to be obtained rather urgently for the final fit-out of the canteen was that of a liquor license. A phone call to the Licensing Court led to the Judge himself very helpfully advising me of the procedure for lodging an application. It appeared that speed was of the essence if the canteen was to be "wet" when the camp was occupied in January. I contacted the legal firm who acted for the company in Adelaide. The solicitor I was referred to said he was fully committed and it would be some weeks before he could help. He was apparently unused to working weekends. After he was advised of the untrue but convincing likelihood of loss of company business he met with me on the Saturday. Early the next week we submitted the completed application and with minimum further delay and fuss an unrestricted liquor license was issued for the canteen.

George in the meantime had received proposals from catering firms who were competent to operate and manage the camp as a whole. They would in addition to feeding the workforce, look after bookings, cleaning, and running the wet canteen. It was decided to sell beer and soft drinks only with no hard liquor. All sales were to be in containers which limited stock basically to stubbies and cans. More importantly it eliminated the need for expensive glass washing equipment with its high water and power requirement. A small shop in the corner of the bar would have toiletries and other odds and sods for sale.

Fuel supply for vehicles, the power plant and aircraft was attended to by BP. The Joint Venture partner agreed to promptly install the necessary tanks and pumps.

In the two weeks before Christmas things finally started to come together. The catering contractor moved in to prepare the kitchen and bunkhouses for occupation in the New Year. John Lunnay had increased his regular transport service from Adelaide, and Telecom had hooked up two old style public phone boxes they had found for us. The water tanks were all filled with the cleaned Olympic Dam water which was pumped in through the now completed small diameter overland pipeline. It received a dose of chlorine and a final polish through a domestic type sand filter on its arrival. A group of farm type pressure pumps with attached air tanks fed into the camp mains and fire hose reels.

The power plant was now in full operation with the air conditioner and cool room demands providing a steady load.

While all this activity was taking place the drillers in their caravan camp at Olympic Dam had continued on with their drilling.

They received their reward when the time came for them to leave for their Christmas break. Instead of the long road trip the first aeroplanes landed on the new and still a bit rough airstrip to fly them direct to Adelaide.

The catering contractor stayed on ensuring that the camp was stocked up and ready to receive the influx of occupants in the New Year. Brett left for a break across the Tasman and was replaced by Ron Martin, a young Civil Engineer, also from Kinhill.

Chapter 15

THE BUILD UP

Roxby Management Services Pty Ltd, immediately to be referred to as RMS, became operational in January of 1980. It was a subsidiary company of Western Mining with the role of looking after the affairs of the Joint Venture and managing the exploration and investigations of the Olympic Dam mineral discovery. The staff and facilities at the Greenhill Road office and the new Olympic Dam village immediately came under its banner. John Copping arrived on transfer from the WMC Nickel Refinery in Western Australia as the Manager leaving George to concentrate on his Geology.

It was more than the three months for which I had initially agreed to come to Adelaide but nevertheless I returned on January 7th. This time I was engaged by RMS instead of WMC Exploration Division other than for time spent on monitoring the new office construction.

The next day it was off to Olympic Dam for a two night stay in the new camp. The big event on the first night was the opening of the wet canteen. I made a point of catching up with Mick Collins to ask him, Ron and Danny to join me there at opening time. It had already been decided that the canteen would trade only for an hour or so before and after the evening meal. I had to look hard to recognize the earthmovers when they arrived. Instead of a thick coating of dust on hard worn work clothes they were all in Sunday best and immaculately groomed. The canteen was crowded and the one that really stood out was Danny. There was not a hair out of place and the gleam of the whitest of smiles in his black face lit up the room. It was a very successful and enjoyable opening as was the meal in the new mess.

The numbers in the camp were increasing quickly as more and more drills and drillers arrived to speed up the delineation of the mineralisation. Ron Martin was learning his way around and was kept busy as work continued apace on setting up the core farm, erecting the drill workshop, and completing the airstrip construction. Northern Earthmovers who had their caravans set up out of the way in the bush to the east of the Olympic Dam claypan, resisted any move into the new camp. They were now busy placing natural gravel to "all weather" the airstrip. We then had them lined them up to form and gravel the tracks for heavy vehicle access throughout the drilling area. This and the preparation and later rehabilitation of drill sites would keep them occupied for some time. The specialised earthmoving fleet from Andamooka that had helped get the airstrip ready for use so quickly continued to help where suited

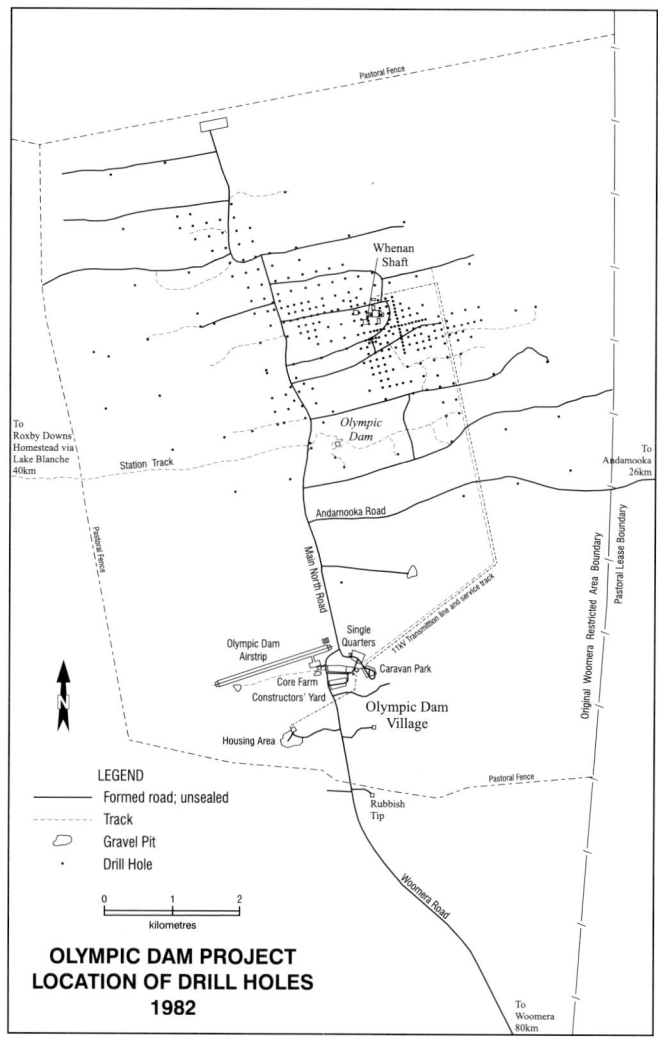

**OLYMPIC DAM PROJECT
LOCATION OF DRILL HOLES
1982**

to our needs. They gravelled vulnerable sections of the road from Andamooka and other sundry works before returning to the opals.

I spent a busy time catching up with the various works and familiarising Ron Martin with the area before returning to Adelaide and settling back in the office. There was plenty to do. The Joint Venture partners were now having technical and management meetings monthly so there were progress reports to prepare for these. There were also the longer term aspects of water supply, access, and communications to look at. John Collins and his drafting staff were handling an increasing amount of geological drafting as the new drills produced core. Never the less John managed to handle the progressive compilation and updating of surface layouts and mapping. The road from the camp to and through the drilling area was now well trafficked and was soon known as 'The Main North Road.' It now extended for four km over the spaced out dunes to the north of Olympic Dam. The interdunal tracks made to access past drill sites branched to the east and west from this 'main' road. The tracks had sensibly been given names by the early drillers and these were faithfully added to the area plans when they were upgraded to roads.

Travelling north from the camp, to the east after the 'Andamooka Road' junction were Greenfield Road, Olympic Dam, then Ironstone Road, Eagle Freeway, Hawk's Nest Road, The Causeway, Salt Bush Road and Drury Lane (East). To the other side was Drury Lane (West), followed by Slack's Track and Brook's Road.

Surprisingly these served the needs without addition over the first years of intensive drilling.

In Adelaide Hansen Yuncken were making excellent progress with the new building WMC were to occupy at 168 Greenhill Road. I had been delegated to keep up with the activity by attending their weekly meetings, requesting any changes that would better serve our occupancy. The building was of two floors above an excavated car park. Each very large floor area was divided into two separate areas with a brick fire wall. The intention was for most of the scattered WMC workforces and RMS to be rehoused in this central office. The big decision was whether to lease all of both floors. This was made after a quick walk through of the large open spaces by those who should know, Arvi Parbo, (later 'Sir Arvi'), and Keith Parry. Keith, then the Director of Operations for the company stayed on in Adelaide to pay a visit to Olympic Dam the next day.

It was an eventful visit for me as that night after dining with Keith, an old neighbour at Kambalda, he put it to me that I rejoin the WMC staff. This was to be as Chief Civil Engineer in the WMC Engineering Services Group (WES) and responsible to Peter Webster the manager. This suited me as I had previously worked for Peter and it meant that I would again be exposed to work on other operations throughout Australia. Although all of the other WES personnel were based in Perth, I would remain in Adelaide in the Roxby office. This made sense as most of the immediate work in my line was in South Australia. In addition to the Olympic Dam project and the new office I was also to become involved in work on the Kingston Coal discovery in the South East of the state. Accompanying Keith to site at 6.30 next morning, after a short sleep on it I accepted his offer of re-appointment.

This was Keith's first visit to the new camp at Olympic Dam. He was quite impressed with the airstrip, thought it cost too much, but still wanted it lengthened. He wished to fly in direct from Perth on his visits, using a Kingair aircraft that required more runway length than the run of the mill twin engined planes. We added this to the list of works.

Ron Martin, who replaced Brett in looking after the ongoing site works for me had, with Kinhill's blessing also agreed to join the WMC staff. Brett returned to Kinhill after his break was and was needed by them on other works. Peter Webster suggested that Ron spend time at Kalgoorlie to become acquainted with the WMC operations on the Goldfields. He was then replaced by Jeremy Folwell a new graduate civil engineer who had grown up at Kambalda where his father was Chief Mining Engineer.

Jeremy took up residence in the camp and soon had plenty to occupy himself, including the installation of small diameter pipelines throughout the drilling area. These would enable the drill rigs to be supplied with saline water for their drilling from a central bore without the need for tankers running all over the site.

The core farm was now filling up with racks of neatly stored trays of drill core. The new sample preparation building as drawn up after my inspection and advice from the staff at Ballarat was being built under contract. George closely followed the planning of this which was in his domain. Particular attention had to be given to the ventilation, dust collection and disposal and storage of wastes. This was because of the uranium content of the drill cores being sawn in half and crushed for assay.

Another visit was made to Woomera to check the final arrangements to purchase water from their supply. The small tank farm, pump and standpipe for loading road tankers at Phillip Ponds was complete, and the small diameter pipeline from the mains at Woomera had sufficient capacity to top up the tanks between loads. It was all ready to go before the cleaned up supply from Olympic Dam gave out exactly on the date we had forecast.

All water to sustain the increasing demand at the camp was now to be tankered from Woomera via Andamooka. The round trip of 284 Km was slow and hard on vehicles making the water very expensive. I did some quick calculations to determine what the savings might be if we had a direct road from Woomera. The round trip distance would be 124 Km less and a reasonable standard of road could result in further savings. There were also indications that the quantity of water required was likely to increase significantly. My findings were that the savings on water cartage alone would be sufficient to justify the cost of a direct road of better standard than the present route. Further significant savings would also be made on the general transport to site which was increasing in quantity daily.

The sample preparation building, when operating, would increase the workforce on site and more drill rigs were arriving. There was now a demand for accommodation for married personnel and we were providing a small caravan park. A larger workforce and more residents meant more water and goods to be carted in on the rough roundabout route.

Chapter 16

THE OLYMPIC DAM TO WOOMERA ROAD

Our whole purpose at Olympic Dam was based on an exploration programme to see whether future mining would be profitable. In those early days there was no certainty of that being the case. The reasons for proposing to spend a lump of the exploration dollars on a new road had to be convincing. A brief report with estimates of costs and savings to be made convinced the Joint Venture Management that we should seek the necessary approvals to build a direct road. They accepted my 'guestimate' of $300,000 to construct the 72 Km length to join the existing road 7 Km out of Woomera. This provided for an 8m wide roadway of raised earth formation on the swale and gibber sections and the ramping up and gravelling over the numerous sand dunes.

I had picked a route for the 36 Km section of road south from Olympic Dam to Purple Downs by stereoscopic examination of the aerial photographs. To minimise interference to the pastoralists this generally followed the boundary between Andamooka and Roxby Downs Stations until it reached the northern fence line of Purple Downs. Continuing on from there it was necessary to stick to high ground on the western ridge line of the large Coolay Lagoon catchment. Tom Allison from Roxby Downs also owned Purple Downs and had sensibly requested that we keep well to the east of the Purple Downs homestead. There we crossed the station track which ran south-east to Arcoona Homestead. To minimize the affects of traffic on his stock we also agreed to fence along sections of the road and to pipe to additional watering points.

The selected alignment to the south then encountered the highest and most formidable dunefield with no easy way around. Beyond that, the last 28 Km of new construction to link with the existing road leading on to Phillip Ponds and Woomera was on open gibber. I had carefully plotted this section of the route on high ground shown on the same contoured map sheet we had compiled when mapping the area in 1947. The area enclosing Woomera was fenced with a locked gate blocking direct access from Phillip Ponds. This meant all of the traffic from Olympic Dam had to take a rough circuitous detour to join the Stuart Highway at Pimba. The next move was to seek the necessary approvals to build the road. Firstly the Commonwealth authorities had to be asked as the alignment was wholly in their "Restricted Area". The line was also not too far distant from one of their installations which might be on the secret list. Then the State authorities had to give their blessing. We were required to be responsible to the State through the Department of

Minerals and Energy. They were supportive of our proposal. We wanted to ensure that our construction was on an alignment suited for permanent access to Olympic Dam if and when mining took place. Then any gravelling and improvements to our initial dirt road would not be wasted.

On March 5th Noel Hiern from the Department and Trevor Code from the Highways office at Port Augusta joined me to travel the route. We left Olympic Dam early, bumping across the gibber and leaving our two wheel tracks over the dunes. Although equipped for a difficult bush bash it was all over early in the afternoon. It was agreed that the line looked reasonable and the next move was for us to submit our proposal for official approval. The Department of the Environment then got into the act. They wanted some detail as to the various affects and impacts of a new road to be submitted by RMS in the form of a "Declaration of Environmental Factors". They agreed that they would check the road corridor for archeological sites as soon as possible. After preparing all of the rest of the submission in the next weeks we waited and waited for their inspection and report. After many weeks we were told that we had to get an independent and approved archeologist to examine and report on our proposal. More weeks passed until his report became available. When it did it was subject to having some artifacts collected that meant a return visit. Then the rules regarding protected plants caused further delay. The whole countryside along the way was liberally dotted with common varieties, and according to the book we would have to seek permits for each and every one we might disturb. This could take months, and was common to any alignment. We sought a blanket approval, producing a booklet with coloured photos to help our supervisors identify and protect these plants where possible.

In the meantime the urgency for a direct road had increased dramatically. In May it had been decided that a 500 m deep shaft would be sunk in the exploration area which would mean a large increase in the workforce, water usage and traffic.

To facilitate planning and recording I had initiated with John Collins' help, a program to prepare standard Australian map sheets for the camp and airstrip areas and to the north of Olympic Dam. These were compiled by contractors and based on the new runs of coloured aerial photography that we now had available. The photogrammetry section of Lands Department had managed to fit these in on one of their flights. They also took a high level photo which we had enlarged in colour to about a metre square. This proved to be an excellent record clearly showing the village buildings,airstrip,the growing road network,drill sites and topography. As an illustration for briefing management and the growing number of interested visitors it was invaluable. More and more copies were obtained as it became a definite status symbol in offices from Perth to London to have one on your wall.

The mapping we required ranged from 1 : 500 for the camp and shaft areas up to 1 : 100,000 for future planning of services and installations. The geologists continued to survey the positions of all their drill holes but the limited maps they produced were not in accordance with the Australian Mapping Grid. When mineralisation was first discovered they had set out a rectangular grid with the east-west axis roughly paralleling the line of sand dunes. The holes were drilled on these grid lines. The five odd years of their geological mapping was all based on this grid and to an arbitrary level datum. They had gone too far to change as far as the geology and underground works were concerned. It was essential that all infrastructure be set out and recorded to Australian standards for levels and mapping co-ordinates. The geologists and mining engineers took some convincing and did not take kindly to having two lots of grid lines shown on the new maps of their territory. I had experienced similar battles on previous mining sites. It was non-negotiable as far as I was concerned as all future leases, licenses and land titles had to be based on standard mapping requirements.

The progress of our mapping was restrained by time and funding. When the shaft site was selected there was a rush to complete large scale contoured maps of the location. Bob Crew, a Mining Engineer with WES in Perth was to look after the shaft sink and was anxious to get started. The site selected happened to be at the edge of a claypan which was obviously the lowest part of what showed up on the photos as a large catchment. The mapping of this area was given priority and in May of 1980 I took the first copies to Perth so planning for the surface works at the shaft site could proceed. The vertical shaft position was rigidly fixed by the underground geology and could not be moved to higher ground. A quick calculation showed that the shaft collar needed to be raised significantly to avoid it becoming a plughole draining storm flow from the claypan and its large catchment into the mine. The rainfall records from the regional homesteads had indicated that the maximum rainfall experienced over 80 to 90 years had been just over 100mm in any month. Playing it safe we raised the collar to take into account twice this which as it happened in the future was just as well.

The sitings of the shaft, winder and compressor room, workshop, change rooms, office, store and storeyard, power plant, hard standings and roads were added to the drawing. It was apparent that most of the claypan also required filling to a depth of well over a metre. To obtain the fill we would excavate a large dam on each side of the claypan to intercept and store storm water runoff. The wheels were set in motion to obtain the required approvals for the shaft, to start preparing the site and getting all the other works that were now required under way.

We added more bunkhouses and extended the mess and canteen at the single quarters and installed an additional and much larger squatters tank to boost the water in storage. More sites with en-suite facilities were added to the caravan park and a recreation hall and tennis courts built. The airstrip was extended in length to 1600 metres and the warm up area at the eastern end and the terminal area sealed to reduce the dust from aircraft movements. The management committee were also persuaded to provide funds for sealing the roads throughout the camp, the van park and alongside the power plant, workshop and core farm.

It was when the proposal for the shaft sink was announced that Aboriginal sites started to appear here and there around the exploration area and near the shaft site. This caused some delays but eventually the approvals to go ahead with the shaft were advised. From then on site clearances became an important factor when works in new areas were being considered.

A reliable and continuous supply of concrete and crushed stone products would now be required for the shaft lining, machinery foundations, buildings and surface works. We invited and received proposals to establish a quarry, crushing, and concrete plant on site. The Readymix group were selected and established a quarry a few km to the north and a concrete plant near the shaft site. The earthmovers were soon busy again excavating large stormwater collection pondages adjacent to the claypan. The excavated material was carted to and used as filling to build up around the shaft site to well above likely flood levels. The fill had to be thoroughly consolidated to carry the heavily loaded foundations of the shaft headframe and plant.

As soon as the earthworks were complete, specialists from the Perth engineering group had contracts under way for the construction of the massive concrete foundations for the shaft collar, headframe and winder. Bill Rymill, a civil engineer from Perth made regular visits to keep the work on track and shortly after relocated to Adelaide. Bob Crew, also at Perth office, was preparing the contract documents for sinking the shaft. Tenders were called and the contract let to Roberts Construction. Bob, who was in overall charge of the shaft works, insisted on having Harry Rymer, a very experienced shaft sinker from the Western Australian Goldfields on site to supervise the contractor.

We were to provide houses for Harry, the shaft contractor, and for George White who would be in charge of activities at Olympic Dam. Also to be housed were the growing number of senior supervisors including those who had been in caravans for some time. The sinking of the shaft was a prelude to bulk sampling of ore and detailed metallurgical testing, and if this was to be carried out at Olympic Dam there could be a need for even more houses. I selected a suitable site for the housing settlement a kilometre or so to the south west of the camp, alongside a small claypan and among stands of large myalls.

I then enlisted the help of another specialist to help with specifying and supervising the building of transportable houses that would stand the rough trip to site. We also wanted to ensure that they could readily be relocated later, either to the future project town or another operation and so retain their full value. Jock Douglas, a wee Scotsman and competent builder had worked with me at Kambalda and on and off ever since. Although small in stature he was like a terrier with a bone in ensuring he received the standards of work he required. The housing suppliers soon learnt to supply as he directed and he earned their solid respect. Over the years we had overcame the worst aspects of the standard transportables by specifying total steel framing with upmarket windows and using compressed cement sheet floors to reduce noise transmission. We produced and progressively modified our own floor plans which were well received by the later occupiers. The houses were fully insulated, had ducted evaporative cooling and room heaters. Care was taken to ensure access to the site and all installations were set out to retain the myalls and other vegetation where possible.

All of this impending activity was making a start on and the speedy completion of the road south essential. It was October when limited approval was finally received. It only covered the section from Olympic Dam to the half way mark at Purple Downs.

We were required to have further studies carried out by an anthropologist on the route from there to the south.

We lost no time in getting started from the Olympic Dam end. Jed Folwell had already pegged the centre line for some distance. The road was set out with high speed value curves so it would be acceptable as a permanent alignment and as a firm base for future upgrading. It would be built by carting in earth fill from off road borrow pits so that it sat above the natural surface other than at dune crossings.

Once again the fleet of earthmoving gear that had appeared from among the mullock heaps at Andamooka was mobilised. A grader lightly cleared the way ahead and spread the dirt brought in by a pair of scrapers. A dozer helped the scrapers load and also pushed sand to ramp over the dunes. A loader and trucks helped out by carting in gravelly material to surface the sand at the dune crossings. One deficiency was enough water tankers to dampen the fill and tracks to the borrow pits to reduce the ever present dust. The dune crossings being up and over were a traffic hazard because of very limited sight distance. We increased the width of the road at these crests to enable opposing traffic to pass safely. Signs advising of the hazard were erected at the start of each dunefield to alert those not familiar with the conditions.

Tom Allison also helped out by flying me along the route at low level in his small aircraft. His local knowledge enabled him to indicate all the prospective areas for obtaining gravel surfacing. The Highways Department who we kept up to date with our plans and progress continued to take an interest. They assured us that our lightly built road would surely become corrugated similar to most of their outback dirt roads. We pressed on regardless with Jed looking after the roadworks as well as the extensions to the single quarters, the caravan park, and the preparation of the housing area.

The road construction was nearing Purple Downs when approval to continue southward was received. It was very heavy going and slow through the last of the dunes with deep cuttings and fills in sand that had to be sheeted with gravel.It was then on to the long stretch of open gibber that seemed endless. The summer was upon us, which together with increasing numbers on site meant more water carting so the pressure was on to get the road open. The heavy lifts of the shaft steelwork and plant were also increasing and the transport operators were getting anxious to have access to the direct road.

There had been ample gravel in the borrow pits to ensure an all weather surface through the dunes and interdunal swales. This changed dramatically on the gibber tableland when all that was available under the thin layer of sunburnt stones was dense clay. As the clay formation took shape we continued watering with the saline water from a nearby salt lake, This held the surface together which could be graded as smooth as billiard table in the dry and was comfortable at highway speeds. When rain inevitably fell all changed. Initially it became extremely slippery. If the rain persisted and long stretches became rutted with twisting wheel marks. As the travellers tried to steer a straight course through the mire mud built up on the wheels and although progress was slow it was still possible.

The new road was finally ready for traffic by March of 1981 cutting 62 Km and at least an hour off the run to Woomera and beyond. The mainly dry road travelled so well that as well as serving Olympic Dam it also became the accepted route to Andamooka. This suited the Highways Department as it largely relieved them of the need to maintain the old road and the constant complaints from the residents at it's condition diminished. The maintenance of the new road to where it joined the original at the Arcoona station turnoff, 7 km from Woomera, remained our responsibility for the years that followed.

Chapter 17

LET'S HAVE A SHAFT

There had also been a lot happening in Adelaide as staff numbers increased and the temporary office filled up. The continued monitoring of the speedy construction of the new building at 168 Greenhill Road had been taking up more of my time. The builders, Hansen Yuncken and their architect were most co-operative making any changes we deemed as necessary such as double glazing to reduce traffic noise along the street frontage. They also agreed to contract to us the fit out and partitioning and carry it out in conjunction with the building works if we provided the detail promptly. This gave us the opportunity of occupation some months earlier and at a competitive cost. I commissioned their architect to help with fitting all the incoming WMC groups in the space available.

It was agreed that Roy Woodall, the Director of Exploration and his highly technical and scientific staff would occupy the front section of the raised ground floor. The rear section of the same floor was allocated to Roy's "Petroleum Division. Both of these groups had their particular requirements for laboratories, dark rooms, fire proof strong rooms and other complicated installations. Roxby Management Services were to occupy the front section of the upper floor. Their needs were more modest but did include a sound proof separately air conditioned conference room and a front row of offices. The rear section of the upper floor was to remain empty for the present. Added to the top of the building was a radio mast designed for communication with the radios used by Exploration field staff in the bush.

We were aware of the building becoming a likely target for the anti-uranium graffiti artists that had previously been attracted to the Marion Road office fence. To help combat this we had the brickwork at the front painted with an anti-graffiti paint. To enhance security and to control access to the building and its sensitive areas the latest in programmable magnetic keys were installed. These let particular people have after hours access to the building. They were also programmed to limit access to the offices and conference rooms where confidential material might be displayed. This was mainly because of the sensitivity of the stock market at the time to any information that could be misinterpreted as good or bad.

Eventually we moved in on July 31st. The next day the top brass from BP London and WMC arrived for the first "Management Meeting" in the new security keyed conference room. The only catch - none of the new fangled

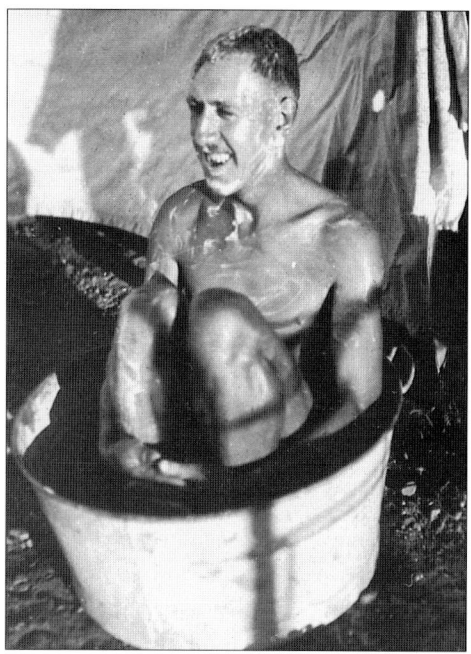

Bathing facilities – Phillip Ponds. 1947 (As displayed at the office in Adelaide in 1980)

micro-chip keys would open the door and let them in. A very tense half hour followed. Our distinguished visitors and WMC Directors busied themselves checking out the offices and the photos of the latest at Olympic Dam as displayed in a photographic exhibition at the reception area. I featured in this without my knowledge. An historical photo of me taking a bath in a tin tub at Phillip Ponds in 1947 had been quietly enlarged and pinned up for all to see.

The caption "Bathing facilities - Olympic Dam."

The weeks that followed saw the settling in at the new office with the staff numbers on the RMS floor increasing dramatically. Henry Muller, moved in as Chief Metallurgist to determine how to best treat the ore that would come from any mining and obtain the maximum product. This had to be done on a very small scale as until the shaft produced bulk samples there was only core from the diamond drilling for testing. Peter Batten the Chief Accountant also moved in to a front row office as did Geoff Witham a Legal Officer from WMC. The geological and general drafting workforce directed by John Collins was also increasing, as was the space required to store their drawings. When two Mining Engineers arrived we had to start partitioning the rear section of the office to house them and also Bill Rymill who moved over from Perth.

We were also catering for regular visits from Dave Thomas and his staff from the Joint Venture Partner, BP, who were to maintain a check on our work and planning on a continuing basis. A great deal of the engineering for the mine electrical and mechanical services was now being carried out by the WMC staff from Perth. The power demand for the initial works at the shaft was to be provided by the village power plant. Much larger diesel-electric sets would be installed at the shaft site when available. These would be needed to operate the shaft winder and to meet the heavy surface and underground power demands. They would also have capacity to more economically supply the growing power load at the village leaving the plant there on standby. The urgent need was for a 11kv high tension power line linking the two sites to get the works on the shaft-sink under way. It also highlighted the need for long term land use planning. To maintain more than ample height clearances on the airstrip approach we located the new power line well to the east of the camp. A section of the road from the camp to the mine would also need to be rebuilt to the east at a later stage as it was too close to the end of the runway.

A single lane gravelled track was built along the route of the power line to provide access for construction and maintenance. A 20 pair cable for communications and to link the two power plants was ploughed in on the same route. It became standard procedure to provide this narrow traffic way, gravelled as necessary, along all routes for future services. We could then insist that all construction and other vehicles travel on the defined track at all times. The result was minimum disturbance to the surrounding vegetation and less dusty conditions. Previously when the alignment for the new road to Woomera was proposed a multitude of authorities found the need to inspect the route. We had not been permitted to mark it other than with isolated survey pegs. The nett result that most of those on their inspections lost sight of the pegs and criss crossed the alignment with indented wheel tracks causing considerable avoidable damage to the vegetation. A review of procedures with officers of the Mines and Environment Departments led to approval to mark future proposed routes with a single width dozer shave.

The power line to the mine and another to the housing area were soon completed and sites prepared for the first buildings at each location. An increasing number of staff and important visitors were now needing to stay in the camp. A solidly built and specially designed staff unit was provided to meet the need and immediately designated "The Hilton." Additional "Atco" sections were added to the kitchen, mess, and wet canteen, and as bunkhouses in the camp. The caravan park kept growing and additional families settled in. The transportable sections of the first houses were brought in on the new road and assembled.

George and Gillian White and family occupied the first house in January 1981. The settlement was becoming a real village.

George now assumed direct responsibility for all activities on site in a resident capacity. We had installed some Atco units at the core farm as a geological office. This with the now busy sample preparation building, the adjacent drill workshop and store made this area a centre of activity.

The growth of the van park and occupation of additional houses brought the days of the all male workforce to an end. Women were filling positions in sample preparation, in offices and in the camp to the extent that positions were found for all. Everyone was entitled to eat at the mess at a fixed moderate cost which was a great help to those involved. Provision was made in the staff

Olympic Dam – Development – mid-1980. A good season

Olympic Dam – Development – 1982. Start of a drought

Camp, Van Park, and Core Farm

The Road to Whenan Shaft,
1980 before the Headframe and Start of The Shaftsinking

accommodation units for single female staff and visitors. The other civilizing change was the increasing number of children and there was soon the need to upgrade the road to Andamooka for school bus travel. The school there was extremely well equipped under the special provisions for isolated areas and welcomed the influx of new pupils. A disused transportable building from a southern area was also landed on a site at the entrance to the camp and used for the pre-schoolers. The facilities for the growing population were added to with the installation of transportable units fitted out as a medical, dental and ambulance centre. A resident Nursing Sister and First Aid Attendant looked after the day to day problems of the community and were backed up with regular visits from the Flying Doctor service and a dentist. The company had a community hall built for screening of movies and playing badminton. Then came a small swimming pool, and a T.V. transmitter was installed. Gaining approval for the latter had stalled with the bureaucrats in Canberra with no end in sight so we went ahead anyway. A mesh-dish receiving antenna on the sand dune near the camp and an imported low output transmitter enabled reception from a satellite of the ABC transmission from Newcastle. The Canberra regulators finally accepted it as an experimental installation.

All proceeds from the canteen were used for community facilities and helped enthusiasts build a squash court. The wet canteen, during opening hours, was by now a social centre and the floor space although enlarged was at a premium. Long term regulars had their special positions. Dasher (Ian) Davey was a fixture against the wall at the end of the short side of the bar loudly holding court with authority. A sign on the wall reserved and proclaimed it as 'Dashers Corner'. A night out, particularly if entertaining visitors was to go to Andamooka for a meal at the Pub. This usually resulted in a late return, the visitors somewhat the worse for wear and quieter than usual the next day. There was no shortage of social life in an apparently self contained and settled community. The new road enabled an interchange for sporting and shopping trips to Woomera. Port Augusta also became in reach for a day or weekend visit.

As work at the shaft site gained momentum more and more contractors arrived in the camp. The airstrip was becoming busier with company and contractor charter aircraft making day trips most days of the week and the odd tourist dropping in for a look. Their refueling needs were attended to by Bluey Lavrick. Bluey and his wife Jan moved to Olympic Dam from the W.A. Goldfields in early 1980. They had the contract for the mechanical maintenance and repairs at the drill workshop. Bluey also had a small single engined aircraft parked at the terminal and kept a close eye on the place. There was always plenty of willing helpers to run out flares along each side of the runway on those occasions when a Flying Doctor aircraft was needed for an emergency at night. We had to run a fence around the perimeter of the airstrip as the good season had attracted emus and kangaroos causing a hazard.

Another problem that arose because of the lush growth was rabbits in plague numbers. Even in the daytime they were literally under your feet when walking from the airstrip to the camp. We made the place available to a group of rabbit shooters giving them power for their cool room. Soon they were reputed to be accounting for 200 each a night but made little impression on the numbers at loose. Being brought up in the bush with ferrets I was amazed when asked by George, who came from inner London how to kill a rabbit. After a demonstration and advice on preparation for the table no doubt they were able to enjoy them on their menu. Time cured the plague at the expense of everything edible at ground level and completely bare dirt over the whole area.

The cattle had long since been removed from the Olympic Dam watering point by arrangement with Tom Allison. This was not only to let the vegetation recover over their grazing radius of six to eight miles. It was also to prevent them having access to the drill sites where cuttings from the uranium containing ore were contained in the mud ponds pending storage. We added to the station fencing to ensure that our area was secure.

The decision to sink the shaft raised the need to gather all the data we could on the saline water aquifer in the quartzites underlying the surface limestone. Cec Forbes of Australian Groundwater Consultants (AGC) was asked to assess, monitor and advise on this as well as managing an initial exploration programme for long term project water supply.

He arranged for Ian Rowan, one of their staff to carry out the field work. The aquifer static water levels were obtainable right throughout the mineralised area and beyond at each drill site. It was simply a matter of removing the cap on the drill casing and measuring the depth to water in the hole. The ground levels at each were already recorded so we soon had an accurate contour plan of water levels in the aquifer. Although there were some humps and hollows indicating local discontinuities, the general grade was towards the east and Lake Torrens. We were now in a good position to monitor changes in the levels when the shaft sink penetrated the aquifer and pumping from the excavation became necessary.

Bob Crew was active in progressing the works at the shaft site following environmental approvals and had relocated from Perth to Adelaide. As soon as the surface facilities and preparatory works were ready the shaft sinking proper commenced and the hole was lined with concrete from the surface to the start of solid rock.

To formalise the occupancy of the land subject to the Joint Venture activities at Olympic Dam the State had granted an Exploration License of about 140 square kilometres. This extended from well north of the known mineralised zone to south of the camp. Although primarily on Roxby Downs

Olympic Dam Village Housing Area – 1984

School Bus on Road to Andamooke – 1983

it overlapped onto the Andamooka station by about two kilometres. Within the bounds of this license was now a Retention Lease around the shaft area, a Mining Lease around the quarry producing crushed rock, and a Miscellaneous Purposes Licences covering the camp, airstrip and surrounds.

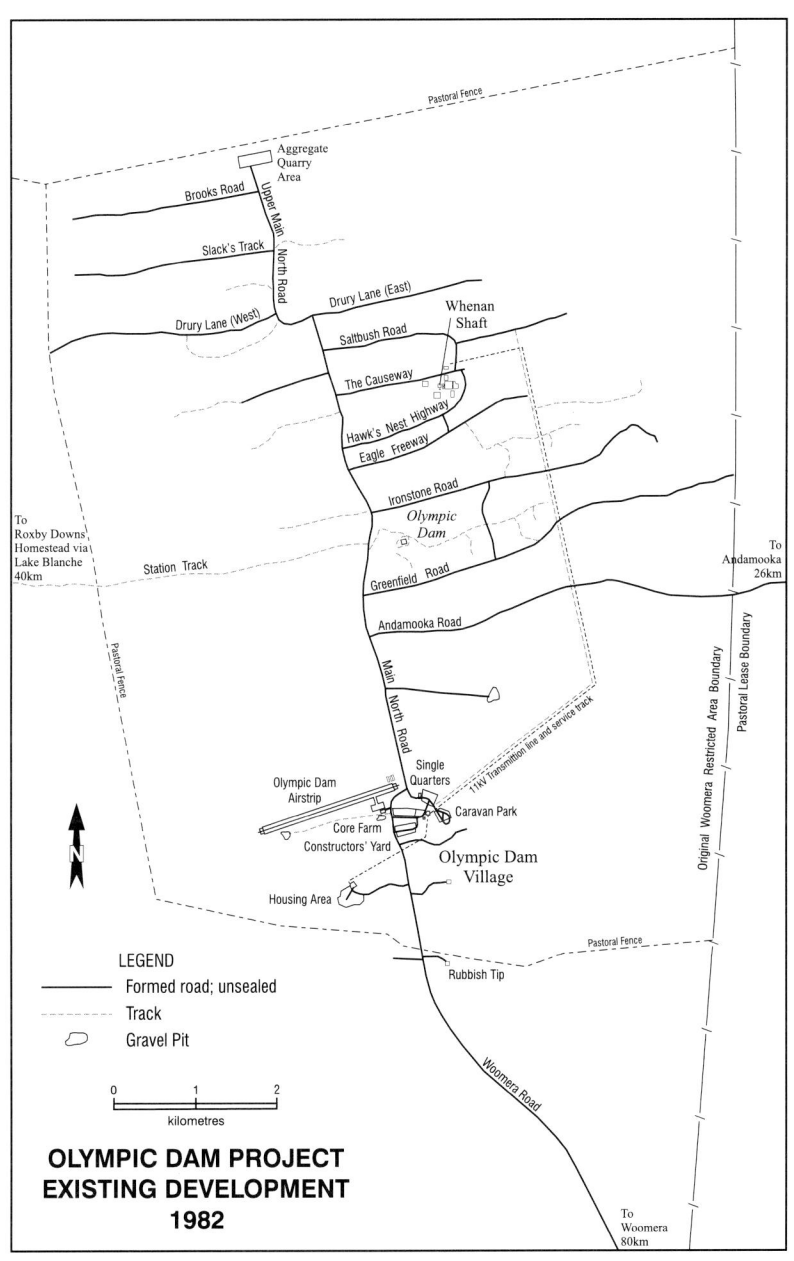

LEGEND

— Formed road; unsealed

---- Track

Gravel Pit

0   1   2
kilometres

**OLYMPIC DAM PROJECT
EXISTING DEVELOPMENT
1982**

Chapter 18

THE ENVIRONMENTAL STUDY
and THE INDENTURE

The intensive drilling continued to confirm the potential for the discovery to become a world class mine. However there would be years of very high expenditure to prove if development was technically and financially viable. The Joint Venturers correctly decided that to justify these costs and to continue with the investigations they needed the security of the right to establish a mine under specific conditions.

The first need would be an agreement with the State in the form of an "Indenture." This would need to be ratified by Parliament so as to become law. It's purpose was to set out the rights and obligations of the Joint Venturers and the State as the exploration and evaluations continued and in the event of a mine being developed.

Secondly the full environmental approvals of both the State and the Commonwealth needed to be obtained, particularly those aspects relating to the mining, production, handling and sales of the metals including uranium. That required the preparation of a detailed "Environmental Impact Statement" covering all issues including the approval process.

Geoff Witham, a WMC legal eagle, already settled in at the RMS office, busied himself negotiating the clauses of the Indenture with the State. At the same time Don Aldridge took up residence to co-ordinate the collection of input from the RMS staff for the EIS. The Department of Mines also boosted their staff dealing with the Olympic Dam activities. Peter Hill, a mining engineer with extensive experience with the active Department of State Development in W.A. was the first appointment. Peter would co-ordinate the activities of the various State authorities that would have responsibilities in any development. Then came John Harris with municipal engineering and secretarial qualifications to deal with matters affecting the development of the town that would be needed if mining started.

It had been decided that the Indenture should deal with a Mine to produce up to 6.5 million tonnes of ore a year, and a Metallurgical Plant turning out up to 150,000 tonnes of copper metal. An associated and contentious product would be uranium oxide at a rate up to 3,000 tonnes. Other metals extracted would be up to 3,400 kilograms of gold and 23,000 kilograms of silver. Prior to investing in such a project the company would need security of tenure of the lands required for the mine, plant, services and a town. They also had to have rights to ample water and power supplies and to know the basis of charging for these and the royalties the State would receive for the mine production.

The preparation of the EIS was entrusted to Kinhill in conjunction with U.S.A. based consultants, Stearns Rogers. The scope of the project to be covered by the EIS was to conform to that being considered in the Indenture.

Dr Brian Jenkins, a young Kinhill staff member was given the task of the day to day management of the wide range of consultants they engaged to contribute to the EIS. He also had the more difficult task of extracting from those of us in the RMS office and our consultants the precise details of what the project would consist of. Brian was well versed in the sciences but not quite so well in the practicalities of engineering. Whereas we "had been there and done that" as far as a lot of our planning of works was concerned, Brian wanted to hear of all the alternatives there were and why we proposed what we did. At times it seemed that we were giving a course in basic engineering.

In January of 1981 the first of the innumerable meetings needed to progress the gathering of the mass of information for preparation of the EIS began. I knew what was to come having been involved in a similar exercise in the West. My involvement was to provide in great detail all aspects as to how we proposed to develop a township, water supplies, waste disposal facilities, tailings storages, roads, and the airstrip. There were other sundry details such as stormwater drainage, mapping, and tele-communications. The proposed alignments of and description and sizes of incoming services including power lines from both Woomera and Port Augusta had to be defined. We needed to determine where the water supplies would be sourced, and if as seemed likely, it was from the Artesian Basin to the north, the locations and pipeline routes. Our proposals for developing a township were also to be provided in detail, including site selection, planning, types of accommodation, facilities and administration. Similar intensive detailed information was required in regard to the geology, mining and metallurgy, as well as the plant and associated buildings. This meant that the Adelaide office filled up quickly with specialist WMC staff to deal with all this. At the same time as satisfying the needs of the consultants, we were all deeply involved in continuing project studies. These were to define the needs and estimate the capital and operating costs of an initial project and increasing production at various rates. Peter Grief and Tony Owens, the Mining Engineers who had now settled in at the rear of the office joined Bob Crew in defining the input on mine development. Mike Softley, a Mechanical Engineer moved over from the WES Perth office and settled in beside them. Mike, a specialist on mining and metallurgical plant installations, had been with WMC since before Kambalda was discovered. He joined with Henry Muller and his offsiders in providing details of what the metallurgical plant would consist of. Henry also brought his wide experience to focus on the copper and uranium extraction plant and processing. Ross Muller, (no relation), joined him with specialised smelter operating experience.

Don Aldridge faced up to co-ordinate the studies and to ensure that the environmental inputs, meetings, and managerial needs were met. John Collins and his drafting staff were extended to the limit endeavouring to meet the needs of the many consultants for maps and drawings showing the precise locations and detail of all proposed works. At the same time there was an increasing load of geological drafting as well as engineering drawings.

There was also increasing activity on site. The shaft contractors had started on three shifts with a 30 man workforce as soon as the winder and headframe were installed. The concrete plant was in operation and the drills in full production. The EIS consultants were there in numbers to prepare their assessments of the existing conditions and what affects would result from development.

We continued to add to the camp and caravan park. Bob Crew and family moved in to one of the additional houses added to the settlement. Bob succeeded George as Manager on site and we designed a transportable administration office. This was solidly built in similar construction to the houses. It was placed on a site alongside the airstrip terminal which proved very convenient. Adelaide staff were regular visitors for site meetings in the conference room in the building and only had a 50 metre walk on arrival. The office soon filled up and extensions were necessary as other staff were appointed.

Chapter 19

WATER and THE GREAT ARTESIAN BASIN

Among the major matters yet to be determined for EIS input and Indenture considerations were the source and development details for the project water supplies. We already had AGC check out the regional water resources and had eliminated the possibility of finding reasonable quality groundwater in quantity. The option of developing a storage to collect storm runoff in the Flinders Ranges on the far side of lake Torrens was also eliminated. That left only the high cost limited supply of Murray River water from Port Augusta, a higher cost direct pipeline from the river at Morgan, or from a borefield or fields in the Great Artesian Basin.

Analysis of the water from existing bores in the GAB showed that desalination would be needed to produce potable quality supplies. However in its raw state with the saline reject from a desalination plant added it should be suitable for use in the metallurgical plant. AGC were instructed to get moving with the drilling and testing of bores in prospective areas near the southern boundary of the artesian basin and as close to the mine as reasonable. They obtained the necessary permits to drill and called for prices from licensed drillers.

Discussions were held with the hydrogeological specialists in both the Mines and Water Supply Departments, and records of existing producing bores studied. The nearest potential area was centred on a pastoral bore on Stuart Creek Station about 15 km to the south of Lake Eyre South. Known as New Years Gift Bore it was about 110 km north of the mine and had been flowing freely to the surface under artesian pressure for many years. The overflow from a stock trough receiving the discharge had established a wetland on the dry barren gibber. This expansive area, lush with a dense growth of rushes had become a haven for rabbits, kangaroos, and numerous birds including brolgas. The signs of this region once being part of an ancient seashore were indicated by the fossilised mussel and other shells to be found on the surface.

Seismic soundings followed by drilling and testing defined the characteristics of the aquifer in the area and its potential for production. Initially we drilled five test holes at distances of five to ten km around New Years Gift bore. They all intersected about fifteen metres of aquifer at depths of 50 to 70 metres. Each bore flowed freely with a shut in head at the surface of about 20 metres. The water had a temperature of around 25 degrees and 2500 parts per million of dissolved solids. We designated this area as potential "Borefield A".

The drilling here showed that we were close to the southern boundary of the GAB and that there would be a limit on the safe yield that could be expected from this area. Outcropping basement rocks to the east indicated that flow curved into the area from under Lake Eyre so there were limits to extending a borefield in that direction. Borefield A would probably provide sufficient water for an initial supply at the shortest distance.

To supply the very large mine being considered in the EIS we would have to draw additional supplies from a "Borefield B". That would have to be well beyond the regional high point and landmark of Hermit Hill, situated 15 km to the east of New Years Gift bore and just to the north of the Oodnadatta track.

Hermit Hill, 60 km to the west of Marree is on a ridge of the ancient basement rocks that form a southerly boundary to the artesian basin sedimentary aquifers. Along this edge where these aquifers become shallow, seepage flows to the surface have occurred over the milleniums. A myriad of unique mound springs, seepages, and wetlands has resulted, spaced at intervals along the fringes of the basin over thousands of kilometres. In our arid area of interest the seepages fed small permanent wetlands or into the sandy stream beds leading to Lake Eyre South.

AGC had compiled a comprehensive listing of all of the pastoral bores and springs in the region together with details of the flow from each and whether they were free flowing or controlled. They also had collected water samples to determine the water quality from each.

In the area 40 kilometres to the north east of Hermit Hill was "Crows Nest" bore which in past years flowed freely to the surface at over six million litres a day. It had recently been controlled and throttled back to about one tenth of that. A further 20 km to the east was "Big Bore", which had flowed freely at about seven million litres per day since it was drilled in the 1920's. It was obvious that these had tapped a very high yielding artesian basin aquifer and that the particular area had the potential to provide the additional water for a major project. The distance for a pipeline to the mine would be about 150 km but there was no chance of finding a closer source. The costs of developing a borefield would be very high as seismic testing indicated depths to the bedrock at about 380 metres at Crows Nest. This was an increase from 100 metres at the Marree to Oodnadatta road 30 km to the south.

We designated this area as "Borefield B". To thoroughly drill, test and prove this deep aquifer as able to supply the project's needs would be extremely costly. The Joint Venturers were reluctant to provide the funds, as there was no certainty at the time that they would ever have a producing mine. They had to be content with my assurances that there, or at the worst, further east, the water could be found.

They did however approve of a seismic traverse and the drilling of two test holes 5 km and 20 km to the south of the high yielding Crows Nest bore where the aquifer was much shallower. This was to check if the very productive aquifer extended in that direction which was closer to the mine. It did not. The holes were still valuable in adding monitoring positions for measuring variations in the artesian pressures.

The high pressure of the water in the artesian basin meant that any drilling had to be carried out under controlled conditions by specially licensed drillers. Surface casing had to be cemented in to sufficient depth and fitted with valves to contain the high shut in pressure of the bore. The drilling was then carried out with heavily weighted mud circulating in the hole to prevent a blow out when the aquifer was reached.

The analysis of the initial flow tests from the exploration bores at Borefield A showed that this area would safely yield at least 6 megalitres per day. The quantity would be sufficient to meet the needs of the construction and start up phases of the project. It would be a relatively economical development because of the shallow aquifer and a delivery pipeline distance of about 110 kilometres. To meet the far larger requirements for a major project Borefield B would have to be developed in the area of Crows Nest and Big bore. This being about 45 km to the north west of Marree and 150 km from the mine.

AGC were regularly monitoring the flows, pressures, temperatures and quality in all of the accessible bores and many of the natural springs in the region. Analysis of the aquifer water quality from throughout confirmed that all samples were too brackish to meet the standards for potable use. We contacted suppliers of desalination plants to determine what processes would produce potable water from these. Henry Muller was also very interested to see if there were any obvious salts in the samples that might cause metallurgical problems.

There were no roads to the north of Olympic Dam leading to the area where we started drilling other than well out of the way narrow station tracks. The drill and drillers caravans had been brought in on the Oodnadatta Track from Maree. Geoff Whitham and I had visited the Adelaide office of the Kidman Pastoral Company who controlled the Stuart Creek Station prior to the start of drilling. They were very supportive and paved the way to land at the station airstrip when visiting the work. Ian Rowan, the AGC hydrogeologist in charge of the drilling would pick us up for inspections of the drill sites and the country to the north west.

On the first trip we also visited Maree, and continued on to Muloorina Station, calling at "Big Bore" on the way. It was amazing to see a pipe sticking out of the ground in this bare, barren, treeless landscape gushing out hot, clear water. Also that it had been doing so continuously at close to two million gallons a day since the bore was drilled over 50 years ago. The discharge from the bore splashed into a steaming pond then flowed in an open drain for miles to a small lake at Muloorina homestead. We had arranged for the aircraft to be waiting at the station airstrip when we reached there so after a quick visit we were able to return direct to Olympic Dam.

The findings from our investigations and drilling in the GAB were becoming important for several reasons. The works required and the costs of providing an assured supply of both potable and process quality water were needed firstly for the continuing studies we were doing. The quantities of water required varied according to the mine production being considered and ranged upwards from six to in excess of thirty megalitres a day. (I was slowly coming to terms with metrification.)

Secondly the EIS consultants needed to know from which areas we were going to to develop our water supplies and what affects the extraction of water might have. Of particular interest were the the numerous mound springs, their flows, and the associated aquatic life and vegetation. These were becoming a very significant issue in the studies they were carrying out.

There was also the matter of the clauses relating to water supplies in the drafts of the Indenture. These included the rights and obligations in establishing borefields and supplying the mine from the GAB. It was left to Cec Forbes and I to advise, and with Geoff Witham, negotiate conditions with the authorities and the Minister with responsibility for the Indenture. A critical and responsible clause required the Joint Venturers to limit the reductions in pressure in the aquifer at the boundaries of a designated area surrounding any borefield.

We needed to extend our understanding of the Borefield A aquifers and to obtain the information to accurately predict the affects of the extraction of water from the area. AGC advised that a long term flow test from a production bore to be constructed in the area of the borefield was necessary. The flow test together with measurements of resulting drawdowns in surrounding monitoring bores would provide data to enable the computer modelling of borefield behavior. As well as our existing exploration holes we would need to drill more monitoring bores extending up to 15 km to the north west near the South Lake Eyre shoreline. We also elected to drill several holes at a safe distance out from the Hermit Hill springs to allow for monitoring of existing aquifer pressures and any changes that might occur.

A site about 5 km north of New Years Gift bore and on the east side of the wide dry channel of Gregory Creek was selected from a seismic traverse and location as suitable for the long term flow test. Prior to constructing a production bore a small diameter hole is first drilled and the depth, thickness and physical characteristics of the aquifer rock determined. These together with results from flowing the hole at various rates provide the information to design the production bore. This entails prescribing the length, mesh size and depth of the stainless steel screen, and size of the screenings for the outer packing.

The production bore is then drilled a short distance away from the test hole which is retained and fitted with pressure gauges. This enabled the recording of the progressive pressure drawdowns as the flow testing proceeded. The bores numbered 'Gab 6' and '6A' were drilled, early in 1983. The borehead valve was adjusted to give a discharge of about two megalitres a day which flowed freely out on to the bare stoney tableland and into the sandy bed of Gregory Creek. The pressures at all the near and far monitoring bores were recorded monthly and the results then tabulated and with a map of the region showing contours of drawdown advised to the Mines and Water Supply Departments. The computer modelling of results from the first months of the test showed that Borefield A would safely yield 8 to 9 and possibly more megalitres a day in the long term. It also indicated that the expected drawdowns should not adversely affect the sensitive mound springs.

By this time we had a narrow, one dozer width track direct from Olympic Dam to Borefied A which linked with a station track from New Years Gift to Bopeechee on the Oodnadatta Track.

This came about when the EIS consultants required us to accurately define the future pipeline corridor. The various specialists needed to examine it closely to map the terrain, vegetation, fauna and features on which the pipeline might impact. I had selected what seemed the best route after close examination of all the available mapping and aerial photos. We obtained approval from the State to survey the alignment and define it with a one blade width pass with a light dozer.

An archaeological survey along the line found the remnants of old aboriginal camp sites with their stone chips in sand dunes as expected. These are common throughout the area and nothing was found that warranted further action. An anthropologist experienced in the area also checked out that the route was clear of mythological and significant sites. The alignment was pegged by a surveyor who was doing some drafting at the RMS office and was keen to go bush. We equipped him with a 4 wd vehicle, camping gear, a chainman, and a light crawler tractor. He was duly instructed to generally follow the line as marked on the maps and photos but to make minor deviations if desirable. A narrow cleared line resulted, up and over the dunes, across the barren tablelands, and around the dry lakes and canegrass swamps. This soon became the dry weather only and regular access to the activity at the artesian basin.

When flow testing at bore Gab 6 commenced the warm water was directed over a measuring weir where the flow rate could be checked for adjustment if necessary. It then passed through a small shallow pond before spreading over the bare dry gibber tableland and into Gregory Creek. Within a couple of months there was a dense growth of rushes over the area where the flow spread over the gibber. More surprising was the appearance of numerous small fish in the shallow pond. We discounted the local folklore that they came up from deep underground in the aquifer. The only acceptable explanation seemed to be arrival of eggs in mud on waterbirds' feet.

The fish thrived. Some time later the bore flow had to be shut off for a short while to add another fitting to the outlet. The pond emptied fairly rapidly, with the last of the water draining down a plughole that had mysteriously appeared in the dirt floor. All of the fish disappeared down the hole with the last of the water. An hour or so later when the bore was turned on again and the pond filled the fish swam back out of the hole. We refused to surmise what lay underneath.

Chapter 20

THE TOWN and THE INDENTURE

A major component of any project development would be for the Joint Venturers to provide a town to house the large company, government and private workforce that would result. The only developed township in the region was Woomera. This was regarded as too distant and otherwise unsuitable to accommodate the workforce in view of the expected very long lived mine.

In common with all of the recent major mining projects in remote areas it was recognized that the developers would have to bear all or most of the cost. The rules for the town development would be set out in the Indenture and Geoff Witham enlisted my help in getting our ideas together. Discussions with Peter Hill and John Harris from the Mines Department followed leading to a first rough draft of the matters to be considered.

The company accepted that they would have to provide all of the accommodation by way of houses and single quarters for their direct workforce and key contractors such as mess caterers. They also would be responsible for the design and construction of the roads, streets and services and sought to be the planning authority. There were high level negotiations to have the State provide the schools, community and sporting facilities in return for more favourable royalties payable on mine production.

As decision time neared Colin Wise, a WMC Legal Officer who I had worked with on a similar State Agreement in Western Australia joined us to finalize our approach. Colin was now the "General Counsel" for the company and based in Melbourne. Hence now known as "The Wise man from the East." Many of the company built townships built over recent years had remained as private closed towns for years. Others were making only slow progress towards full local government representation and responsibility.

We held a final meeting to determine what our approach should be in progressing to municipal status for our town. I put forward the proposition that it should be a Municipality from the very start - from day 1 of any project development.

Surprisingly after due consideration there was mutual agreement and we started sorting out the details. The Company would have the obligation to present the State with a 'Project Notice,' if and when, after the necessary approvals, they were to proceed with a mining project. Immediately on the service of that notice the municipality would come into being and would be headed by an Administrator appointed by the State. He would then be responsible for the running of the municipality until a Council was elected. This would be after some years when certain revenue and population figures were reached.

In addition to the normal municipal functions the Council would be the responsible authority for water and electricity supply and the sewerage system. The State would be required to make available to the Company all the land needed for development of the town and for the permanent installations at the village.

The Company would be responsible for the town planning, the subdivisional development, and design and construction of roads, and services. It would also be required to supply water and power in bulk to the Council at the townsite at subsidised rates. This would be so consumers charges would be in line with those applying to customers in the settled areas of the state.

On completion of construction of roads and services in distinct areas of the townsite these would be handed over to the Council without charge. The subdivided allotments could be used for Company housing, single quarters or other purposes or made available for sale for private or State use. The land could only be sold at the actual cost of development.

The State would design and finance the education, medical, basic recreational and sporting facilities, as well as police, municipal and government buildings. However the Company would be the construction authority to ensure timing for essential works could be programmed into the overall development, and in particular to give priority where needed, for contractor accommodation and the provision of services.

The State would through the Highways Department contruct and seal a high standard road from Woomera to the mine lease, with connections to the town and airstrip. This would be at Company cost and undertaken when requested by the Company.

Other matters included in the Indenture provided for the granting of a 'Special Mining Lease' over those parts of the Joint Venture Area required for the mine surface facilities, and the provision for incoming services. These to include easements for power lines from Woomera and Port Augusta, water from the borefields and service lines from the mine to the town.

Geoff Witham pursued the finalisation of the drafting of the Indenture with officers of the Crown Law Department and in negotiations with Roger Goldsworthy, the responsible Minister. Finally all conditions were agreed to, and the document was signed by the Government and the Joint Venturers in March of 1982. It now remained to be passed by an Act of Parliament to become law and for the Company to have the rights to proceed.

A great deal of information on what the make up of the town would be was now required for analysis,inclusion, and discussion in the EIS. It was left to me to come up with the details of where the town would be sited, how it would be developed, and what it would contain. I had no doubts as to the location having carefully scanned all possibilities from the air on each of the approaches and take offs on my many visits to Olympic Dam. Driving over the selected site and careful examination of new coloured aerial photographs confirmed it's suitability and attractions. It firstly was alongside the route to the south and civilisation. It was in sand dune country with plentiful tree and shrub cover with myall flats to the west. The best stand of native pine in the region covered the dunes for several kilometres to the north. The distance from the mine, plant and tailings storages, at 15 km was more than sufficient to ensure that there were no risks from radiation or nuisance from noise or dust. The suitable area was also large enough for a major growth. There was still the need to check the three critical necessities for a satisfactory townsite -

These being :- Drainage, Drainage and Drainage.

Unless there was sufficient fall in land across the site, the interdunal corridors and open swales would after heavy rainfalls become a string of shallow lakes. We committed the funds to extend our survey control into the area and to have contoured maps prepared over the extensive area of potential townsite. A close study of these showed that although careful design would be needed the fall in the ground from east to west was ample for good drainage.

It was now a matter of convincing Brian Jenkins and his EIS consultants that this would be the townsite. I was unsuccessful. It was decided that we should have independent consultants prepare a report covering the merits or otherwise of a number of alternative locations, and to then recommend the preferred site. Ullman and Nolan, a prominent Brisbane based group with extensive experience in developing new mining towns throughout the coalfields in Queensland were appointed. They opened an office in Adelaide with a locally reared Civil Engineer, Brian MacKay as manager. Geoff Nolan, a principal of the firm became involved, making several visits to site and inspected and assessed all of the possible locations. Commuting the workforce from Woomera was ruled out by distance and other obvious factors as was Andamooka. That left six options that I had proposed within the desirable distance of ten to twenty kilometres from the mine. These ranged from due west near Lake Blanche arcing around to the 12 mile dam to the east. It was agreed by all that there was no value in looking to the north. Obvious reasons included the increased distance for incoming traffic and services, less tree cover, and the barrier to access of the mining installations.

It had been deemed that the site must be of sufficient area to develop for a population of up to 9000 people and beyond. This and other factors soon eliminated all of the alternative sites and Ullman and Nolan reported accordingly. We had been calling the accepted location the "Axehead" site. An existing station track crossed the Woomera road leading to the Axehead Dam, which was a few kilometres to the west of the townsite. The track gave good access to the northern end of the townsite so we put signs up for the multitude of visitors that came to have a look.

Many "Town" meetings followed particularly with Peter Hill and John Harris from the Mines Department to fine tune the detail in respect to the town and municipality. It was agreed that the Olympic Dam village would remain and be included in the municipality. The camp would be expanded as necessary as the future "construction camp." The industrial sites that were now established south of the core farm to service the shaft sinking would also stay. They would be part of a permanent non-company heavy industrial area clear of the town and the mine. The van park could remain until the town was developed but the houses would be relocated to the town as soon as the first lots were serviced. It took meeting after meeting to iron out the full detail of all this and to agree all the basic standards for the town roads and services.

John Harris with his municipal background as both a Shire Secretary and Engineer had the role of co-ordinating the State input and reaching agreement on all the details. He kept us busy by arranging meetings with the various authorities as well as regular "town" meetings in our office. We had retained Ullman and Nolan to continue as town consultants and to prepare the initial town plans. Geoff Nolan attended a number of these meetings but had a slightly abrasive manner not appreciated by the local officials. He was however excellent in providing learned detail required both for the information of the State authorities and for preparation of the EIS. He was called on to produce our proposals for such things as the "Conceptual town plan", "Land use budget", "Road hierarchy" and similar items apparently essential for the EIS.

Brian Jenkins needed detailed descriptions and illustrations as to exactly how we were to go about creating a town in amongst the sand dunes. Brian MacKay was kept busy preparing plans and illustrating practical and acceptable means of doing this.

In the meantime the sinking of the 'Whenan' shaft' named after the driller of the 'discovery hole was progressing. In 1981 after passing through the surface sediments and about 50 metres of dolomitic limestone it had encountered the 150 metre thick layers of saturated quartzite and sandstone. These housed the saline water aquifer, and in spite of cement grouting ahead of the excavation, there was a substantial inflow into the shaft. This was pumped to the surface and discharged into a large pond near the shaft site to settle out the sludge and rock cuttings. The water was then made available for road watering and used by the surface and underground drills.

The 100 metres or so of impervious shale below the aquifer prevented the water penetrating deeper into the basement rock containing the mineralisation. The nett result was that it was to be a 'dry mine' other than from the inflows where openings to the surface passed through the quartzite/sandstone layer. By the end of 1981 the shaft had reached a depth of 368 metres in the basement rocks. Sinking continued on into 1982 to the main operating level at 420 metres and final depth of 500 metres. The administration office at the airstrip became busier and busier. A new 48 extension P.A.B.X. telephone system was installed using 5 of our 7 exchange lines, Two lines were still being used for public telephones at the camp. We also put the just introduced facsimile machines in there and at Adelaide.

This incurred the wrath of an officious staff member at WMC head office who considered that I had overstepped the mark, and that they should select these items. As the units were leased there were no repercussions. They helped greatly in the constant communications and the speedy passing of urgent drawings between the two offices.

We were barely able to keep up with the information required by the numerous consultants engaged to prepare the EIS. A detailed Land Use Plan covering the area from the mine to the town was constantly revised and updated. Sites for each facility had to be agreed and added, then the alignments for all of the roads and service corridors. Incoming electric power would initially come in on a transmission line feeding from the existing line to Woomera. When the loading on this neared it's capacity a new higher voltage line would be built from Port Augusta. In defining the precise alignments of these it was often necessary to mark them on the ground as well as on available maps. This was to assist the botanists,terrain analysts, anthropologists, archaeologists and sundry others to locate and inspect and assess the likely impacts.

The restraints that were limiting the possible route into the mine area from the south meant providing an alignment for the incoming power lines in a corridor to the west of the townsite. This also meant that they would pass well to the west of the airstrip and give adequate clearances. However the existing airstrip was not looked upon favourably. It was recommended that if the population grew significantly a new aerodrome should be provided to cater for larger jet aircraft. We chose John Connells as consultants to select a site which they did. This overlapped the Andamooka Station boundary about 5 km to the north east of the townsite.

While the environmental consultants were going about their business, we were kept busy answering their endless questions and justifying our proposals. At the same time we were engaged on continuing studies and estimates for varying rates of mine production and growth. 'The December 1981 Development Proposal' demonstrated that based on the present knowledge of the resource and metal prices a viable mining operation could be brought into production.

Chapter 21

THE TESTING TIME

The Indenture details had reached finality when signed in March of 1982. The State Liberal Government had carefully negotiated the detailed clauses that made up the Indenture agreement. Roger Goldsworthy, the Minister for Minerals and Energy, had ensured that it fairly covered the interests of the State. It required the Joint Venturers to proceed promptly with their continued exploration and evaluation and gave them security of tenure. They were to advise the Minister before the end of 1987 if they were proceeding with a project. As well as the rights to extract water from the southern extremity of the Great Artesian Basin the could draw limited amounts from the pipeline at Port Augusta. The borefields, pumps, pipelines, power lines, and other infrastructure were to be provided by the Joint Venturers. They also had to deliver water and power to the Municipality who would be the supply authority within the municipal area. These services would be heavily subsidised by the company.

The Joint Venturers would have to obtain full environmental approvals from the Commonwealth and the State and operate under continuing strict guidelines for managing, monitoring and reporting.

There was a widespread feeling of uncertainty and concern at this stage due to the current political climate. The Indenture agreement as negotiated with the Liberal Government had to be ratified by both houses of the State Parliament to become law and there was no guarantee of this. The numbers were against it in the Upper House where the Government did not have a majority. The Labour Party policy was against any mining of uranium and would vote against the required legislation. This could mean abandoning the project or at the very least deferring it indefinitely. The anti-uranium mining protesters were also voicing their opinions and were receiving plenty of media coverage.

At risk for the Joint Venturers would be the opportunity to develop a producing mine and to even recoup the high expenditure already legally incurred in evaluating it's potential, and presenting it for approval. This was also a worry for most of those in the RMS office in Adelaide who had put so much effort into it and would face unemployment if the legislation to ratify the Indenture was not passed. Even more at risk were the workforce and families settled in houses and caravans at Olympic Dam. Obviously the Joint Venturers could not continue spending tens of millions of dollars without the security provided in the Indenture. It would be pointless continuing to prove vast ore reserves and metallurgical processes, and obtaining environmental approvals if the resource could not be developed. It would mean a loss of employment for a great many of those associated with the project, both directly and indirectly. I was probably immune as I was still involved in work at a number of other areas including the coal discovery at Kingston in the south east of the State.

There was an extremely anxious wait until Friday June 18th when the vital vote was expected and the un-expected happened. When put to the vote the ratifying bill was passed.

Norman Foster, a staunch Labour Party member in the Upper House 'crossed the floor' and voted with the Government giving them the numbers. He had been a long term and dedicated Labour Member and supporter but paid the price by being immediately thrown out of the Party. He was undoubtably sincere in his support for the project but there is the unanswered question as to whether the Party may have been aware of what was to happen.

In any event the news was immediately received with great delight and relief by all concerned. The RMS staff in Adelaide routinely stayed in the office for a few drinks after work on Fridays. There were no class barriers with the gatherings ranging from visiting Company Directors down to the casual cleaners if they arrived early enough. That Friday the assembly was in a mood of joyous celebration and it was a greatly extended get together

We learned that the news was received even more thankfully and enthusiastically at Olympic Dam where they had far more at stake. The wet canteen would have had a bumper night with a lot of toasts to Mr Norman Foster.

The finalisation of the EIS was now tackled with new vigour. The handling and containment of tailings was one contentious subject and the internationally recognized W.L.P.U. Consultants had been engaged for specialist advice. After due consideration we opted for the sub-aerial method of deposition in a large storage surrounded by compacted earthen embankments. The slurried wastes pumped to the area 5km west of the shaft would be discharged into the storage from around the perimeter banks. This would result in sloping beaches draining to a decanting structure in the middle where the surplus water could be piped to a separate evaporation pond.

Other consultants had fastened on to the unique mound springs of the artesian basin as meriting detailed research and investigations. A couple of Atco accommodation units were landed at our test flowing bore Gab 6 so they and our newly appointed staff environmental officers could base themselves there. It was soon to be known universally as the 'Chicken Ranch.'

The very comprehensive and detailed 560 page "Draft Environmental Impact Statement" was published in October of 1982. Kinhill put on a formal dinner to mark the occasion. I was both surprised and thrilled when Malcolm Kinnaird presented me with a personal copy with my name in gold lettering on the cover.

The document was then open for public review and comment. As expected when submissions were received they were heavily oriented towards the 'No Uranium Mining' points of view. There were others with ill informed and critical comments about the affects of the proposed use of the Great Artesian Basin as the source for water supply.

These were all responded to in the prescribed Supplement to the EIS and eventually both State and Federal environmental approvals were received in 1983.

There was no let up on the project studies being carried out by the RMS staff in Adelaide. These covered the various options of mining methods, metallurgical processes, and siting of the ore hoisting and surface installations. Each new or revised layout required a revision of the plans showing roads and services so we had a very active drawing office. John Collins who was appointed as Chief Draftsman shortly after our first visit to the Lands Department to arrange for new aerial photographs shone in that role. He now had a team of draftsmen looking after the continuing geological mapping. There was also the preparation of the multitude of surface maps and plans needed for each study, for the EIS and the updating of the record plans as new works were completed.

Following the completion and fitting out of the shaft in November 1982, the heavy machinery to start the underground development was lowered down in sections. When re-assembled it was put to work excavating drives out from a platform at 420 metres below the surface. As these progressed out for several hundred metres large holes were drilled from the surface to intersect the drive. The drill holes were then reamed out with a 'raise bore' up to two or three metres in diameter and fitted with large intake or extraction fans in accordance with the mine ventilation designs.

The mine drives out from the shaft continued to be dry with the only water inflows coming from where the shaft and raise bore holes passed through the aquifer. All of the drill holes from the past and continuing exploration drilling were supposed to be fitted with plugs below the aquifer. This was to prevent them piping water into the mine as the drives intersected them. Inevitably one was missed and when one of the advancing drives met up with it there was a deluge into the mine from over 300 metres above. It was panic stations on site before the inflow was stopped, and the flood was directed to the mine pump station and pumped to the surface.

The drives from the shaft gradually lengthened and with more ventilation rises passing through the aquifer, the need arose to make provision for a permanent system to dispose of any excess water in an acceptable way. Peter Hill visited mine with me and agreed with my proposal to pipe the surplus out and evaporate it on a 15 hectare claypan about three and a half kilometres to the west of the mine. This was beyond the known limits of mineralisation and in the general area best suited for tailings storages. No time was lost in laying the pipeline along what was to be a future services corridor. We ensured it was large enough to handle additional mine inflows that could be expected as more ventilation shafts were drilled.

There was a further build up of the site workforce as the activities at the mine increased. Staff were also engaged for more intensive environmental monitoring and the continuing collection of baseline information. Bill Chandler, a specialist in radiation monitoring, who had participated in the Yeelirrie studies transferred in from Perth.

Bill Rymill was now looking after the water exploration and the testing at the GAB. This included the drilling of further bores to the north and north west of Borefield A area towards Lake Eyre South to measure pressure changes resulting from the test flow at bore Gab 6. We were building up a great deal of information as to the nature and behavior of the aquifers surrounding our proposed Borefield A. Regular meetings with Don Armstrong, a hydrogeologist with the Mines Department, kept them up to date and gave us his interpretation of our findings. It became clear that the aquifer in the borefield area was in a valley in the bedrock with the flow southwards up the valley from the deeper sections far below Lake Eyre. We also established that the area had the potential to yield more than the 6 megalitres per day as previously indicated.

To provide more information on existing pressures in the aquifer close to the sensitive Hermit Hill mound springs, further holes were drilled as close as we dared to these. It was going to be necessary for us to show that taking water from the borefield would have minimum impact on the springs and their wetlands.

The passing of the Indenture and the EIS having cleared the way ahead the Joint Venturers decided that it would be prudent to prove the metallurgical processes before considering the vast expenditure for a full scale plant and mine. Drives out from the shaft were now producing bulk samples of various types of ore. Previously the only ore available was the small quantities from the diamond drill cores which were only sufficient for bench scale tests. Continuous testing over a period of months would be necessary to confirm the processes best suited to recover the maximum percentage of the copper, uranium, gold and silver in the ore. The alternatives of testing off-site were considered and rejected.

'Big Bore' – Great Artesian Basin. On Muloorina Station to south of Lake Eyre

A decision was made to construct, at Olympic Dam, a Pilot Plant capable of treating 5 tonnes of ore per hour. This meant a new feeling of confidence and a flurry of activity in Adelaide and on site. Mike Softley was to manage the design and the putting together of the plant. There was backup from the engineering group in Perth and regular visits. A plant site about 1 km to the west of Olympic Dam was carefully selected as suitable. It would need to be securely fenced with controlled access due to the uranium content in the ore. As well as the process testing it would give us the opportunity to closely monitor and determine the characteristics and behavior of the plant tailings.

Henry Muller and his metallurgists were adamant that the pilot plant should operate with the same quality water as would be used when the major project came about. That meant carting substantial quantities in from the artesian basin. Calculations indicated that it would be in excess of 120 kilolitres a day. We would have to build a road the 110 km to Gab 6 suitable for heavy tanker travel. The water from the flow test at Gab 6 would be diverted through a large tank and the road tankers or road trains hauling to site filled from this. This was also going to be a golden opportunity to test the desalination of the artesian basin water for potable use as would be the case in the future. Bill Rymill and I soon were gathering information on the options. We selected a site to store the incoming water, and to install the plant, in an open area about 2 km south east of Olympic Dam. This was close to power from the camp to mine power line, and at reasonable distances from the pilot plant, mine, and camp. It was also a logical terminus for the road from the Gab which would skirt around the western side of the mineralised area.

It seemed sensible to install and trial a desalination plant large enough to produce the total potable water needs for the forseeable future. This would include the increasing usage in the village as well as that required at the mine and pilot plant. It would end our reliance on supply from Woomera who already had very little to spare and who had been considerate to our needs over the years. Looking further ahead we included an allowance for the initial requirement at the start of any major project construction. The cartage rate from the borefield would be similar to that from Woomera, and the cost of desalination would be less than the high cost to purchase from the Woomera supply.

WMC had previous experience with desalination using the 'Reverse Osmosis' process at their Nickel Refinery in Western Australia. We were more inclined to an 'Electro-Dialysis- Reversible' (E.D.R.) process and ordered a plant of this type tailored to our needs. The process water to be used in the Pilot Plant would be the raw bore water with the addition of the reject brine from the desalination. This would correspond with that planned when the mine was in full production. The incoming raw water would be stored in an excavated plastic lined pond and a portion used to produce the desalinated potable water to be pumped to tanks at the points of usage. The remainder with the addition of the reject brine from desalination would be pumped to a tank at the Pilot Plant.

Chapter 22

THE BOREFIELD ROAD

We had appointed a Civil Engineering works supervisor in early 1982 when Jed Folwell was transferred back to the West. Frank Boulton and his family moved into a house at Olympic Dam from the Hydro Electric works in N.W. Tasmania after we had convinced him how great it would be. In no time at all he was extremely busy preparing for and servicing additional houses, van sites, camp buildings and routine area maintenance. He also looked after the initial underground services installations and attended to the needs of the groundwater consultants in the borefield area. Now there was the spate of new works brought about by the decision to build the pilot plant and he was also earmarked to supervise the construction of the road to the borefield. The road would follow the dozer width cleared survey track on the route as selected and shown in the approved EIS.

It would be within a 50 metre wide corridor that would also provide for a future power line and pipelines to the borefields.

At the Olympic Dam end there was an existing road branching off to the east about 3 km north of the camp that ended 2 km along at the site for the Desalination Plant. The first section of the Borefield Road construction would start from there and swing around slowly in a long arc to the north, closing in on the Andamooka Station boundary fence. Then it would follow a station track along the fence line to it's northern end where it would cross the 'Dog Fence' at the 16 km mark. It would then continue north for the remaining 90 km to the borefield on the Stuart Creek Station pastoral lease.

The State approval to build the road was advised on June 28th and construction started from the desalination plant site on July 12th. Bill Rymill had contracted Brambles Ltd Whyalla office to mobilize the plant required and Frank Boulton was supervising the work. A dozer led the way clearing any heavy growth on the 8 metre width, lightly brushing the surface in the swales and forming ramps up and over the dunes. The remaining light vegetation would then be brushed to the side with a grader. Dirt from borrow pits well clear of the road would then be loaded into scrapers, carted and spread along the cleared line resulting in a raised freely draining formation. Dampening down with saline water from the aquifer and final shaping with the grader progressed the road now tagged as 'The Borefield Road' at up to a kilometre a day. The vegetation along the edges remained intact other than at the spaced tracks leading to the borrow pits. We had found over time that our roads built

Borefield Road Construction – On Stuart Creek Station near Borefield A

in this fashion on top of a largely undisturbed surface did not develop the unwanted corrugations so prevalent in the outback.

At the time the State and the Company were being pressured in regard to alleged aboriginal sites being disturbed in the Olympic Dam area. The Kokatha People's Committee, (KPC), based in Port Augusta were claiming that archeological sites in the joint venture area to the north of the mine had been damaged. The Government approval to proceed with the road also included a requirement to salvage the artifacts in an old camp site on a dune near the road about 10 km from the start. The site which had been identified in the EIS inspections was clear of the road alignment and regarded as representative of those common to the region. A Government funded group from the KPC visiting the area from July 12 to 14 in connection with the alleged damage in the mine area were requested to also visit the camp site alongside the road. This they did claiming it to be significant and insisting that it remain undisturbed. They also claimed that there were other sites along the road. The response from RMS was that work on building the road had commenced under the approval granted by the State and would proceed. However we were willing to continue discussions on matters of concern and where practicable, make small scale deviations from the pegged line.

As a follow up the Government agreed to fund the KPC for a further visit to survey the road alignment with RMS and State representatives. This was timed to start on the following Tuesday, July 19th.

Colin Woolard, one of George's geologists had a particular interest in the aboriginal history of the region along with environmental matters. John Copping, noting this, had given Colin the task of investigating the various claims and liasing with the State authorities on them. Colin and I were therefore detailed to represent the Company on the inspection of the road alignment with the aboriginal party.

On Tuesday 19th the dozer clearing the way had progressed 3 km from the start of the road reaching the boundary fence between Roxby and Andamooka stations. It had just encountered a minor stoney ridge which looked to be a good source of gravel and was about to nose into it when the parties for the inspection arrived. It was quite a large group. The State had Mike Madigan and John Wommersley from the Department of the Environment and Ian Kimber, the Mines Inspector from Andamooka in attendance. The aboriginal representatives were Richard Reid, Willie Williams and Max Thomas, They had with them, Chris Charles, a white legal adviser. Colin, Peter Westcott and I were there for the Company. Immediately following the introductions we were told by the aboriginal party that all of this land was theirs, and that where the dozer was about to niggle at the potential gravel was a 'site'. Not wishing to start the inspection adversely we agreed that the low ridge would remain undisturbed. The station track along the fence line that we were developing into the road then skirted a number of small claypans in the hollows between high dunes. Five of these aroused interest and were claimed as 'no go' areas. Protracted on site discussions indicated that if we narrowed the pipeline corridor we could scrape past with some restrictions. It appeared that this short section of dry, sandy, waterless country similar to that for miles around had a site density of one for each square kilometre. They also appeared to be a bit flexible as some areas that were claimed that day seemed to be forgotten later. Frank kept checking with us regularly to ensure he was aware of those locations where some restrictions might apply.

Proceeding north we were requested us to swing the road line further out from the camp site dune where the artifacts were now not to be collected. There appeared to be no further problems on the run to the gate at the dog fence at the 15 km mark. The next 6 km of open gibber was also clear but after ascending a high dune on to an wide elevated stretch of sand we came to a halt. We were looking down at an extensive open area with a few transverse dunes crossing our pegged alignment which continued through in a straight line.

It was quickly pointed out that here was a major problem and there was no way we could continue north on the pegged and defined route. Further discussions were cut short when Max Thomas began feeling unwell and had difficulty breathing. Richard Reid and Willie Williams left to drive him back to his home at Andamooka. In the meantime the dozer clearing for the road had made excellent progress and was pushing a ramp on to the high dune a half a kilometre behind us.

The following day after much deliberation and checking the country for some distance to the north it was agreed that heading off further to the west and skirting around the end of one of the transverse dunes might be acceptable. This only advanced us for about 2 km when we again came to a stop, leaving us a further 3 km short of where it was clear to resume our original route. We then moved north to this point and sought an acceptable line back to the south to close the gap. It was just on dark when there was agreement to the line for most of the closure. It was mainly across a very wide swale and to ensure we had it well defined one vehicle was parked at one end with lights on. The other drove directly in line to the lights from the other end stopping to drive in enough pegs to mark it for the road builders. The darkness and two intervening sand dunes prevented us continuing on to close the remaining 1km gap.

The dozer had reached the end of the gibber plain, and had now finished pushing sand to form the ramp up to the higher ground. It was parked for the night at the start of the proposed deviation at 23 km from Olympic Dam. It was refueled and ready to travel a further 2 km over the gibber to the point all parties had agreed on and to close the remaining 1 km gap the next day.

It was quite late and dark when we bounced along the narrow bulldusty track back to Olympic Dam to find the canteen and the mess closed so we stayed hungry and thirsty. The KPC party advised that they were going on to Andamooka, and would then return to the road line and camp for the night at the dozer.

Thursday morning we reached the parked dozer which was ready to go forward bright and early but there was no sign of the KPC group. I instructed the dozer operator to remain parked while we searched along the line 25 km to the north in case they had made an early start and had gone to have a look up there. For the first time ever in the bush the vehicle I had was fitted with a radio so we called Olympic Dam but there were no messages. After waiting until late morning I instructed a surveyor who was with us for the day to survey and peg the 1 km closure between the north and south agreed lines and to plot the deviation from end to end. At mid-day with no word from the KPC, and after discussion with Mike Madigan, I instructed the dozer driver to start his machine and to blaze the line to the north.

It was an hour later when Brian Motbey, the First Aid Officer at the mine, contacted us by radio to let us know that a message had come in advising that the KPC vehicle had broken down on a station track some distance out of Andamooka. We asked that a mechanic be despatched immediately to get them going.

The dozer was by then had cleared the way ahead and over a small dune. It only had a few hundred metres of gibber and one more dune to cross to close the gap to the line we had pegged in the dark the previous night.

The KPC party at last reached us at the start of the new line and their vehicle again broke down. They were loaded in to our vehicles and we headed along the cleared line. When we reached the dozer, it was half way across the last bit of gibber and had just crossed a small patch of silcrete. The KPC were upset at us having closed the gap without them being present, even though there had been no problem with this area on the previous days inspections. They claimed that the small patch of silcrete immediately behind the dozer was a site and that we had damaged it. Although passed the previous day and not indicated as of any significance it now became very prominent. So much so that the next day it was in headlined in the city newspapers that we had "BULLDOZED A SACRED SITE."

Strangely enough the party on the inspection didn't seem to remain at all upset about it for more than a very short while. They did however make things a little difficult near the end of the deviation where we were to rejoin our original line over nice open country. They now found problems with this and insisted we keep on top of a long eroded ridge of loose sand rather than continue on open, level gibber. Reluctantly we agreed.

On the following day, Friday 22nd, sensitive to the morning's headlines we met up again and continued with the inspection. A small section of white faced breakaway country well clear to the east of our line was indicated as an area to be avoided. We also agreed to deviate our line to keep further away from the shore of a large claypan edged with a stand of large melaleucas.

A kilometre or two on there were four or five very high loose dunes to cross, each separated by narrow gravelly swales. The survey track now consisted of two deep wheel ruts straight up the steep faces of the dunes, on to the crest and down the other side on to the narrow flats. The only way to reach the top was to put the vehicle into four wheel, then take as long a run as possible to build up speed before rushing at the slope. As often as not it stalled or left the wheel ruts swinging madly from side to side until it bogged down. It was then a matter of backing down and trying again. I was travelling with Colin who made me an extremely nervous passenger with his habit of racing flat out at even the smallest dunes.

Eventually we topped the last and highest crest at the 45 km mark and looked down onto a wide extensive basin about 12 km long. It was edged with high sand dunes along its eastern side and had a large dry elongated canegrass swamp of 5 to 6 square km as a core. The open country extended to the west with smaller canegrass swamps scattered through it. Our survey track defining the line of the future road skirted around the western side of and well clear of the swamp. It was on smooth level ground with no obstacles through to the 57 km mark where it entered a very long swale between wide low dunes. The dunes to the south all ran in an east-west direction but from here on they ran in unbroken lines to the north. This suited us well as the road sat comfortably in the swale between two dunes and there were no further dune crossings for the rest of the way to the borefield.

The whole basin we now knew as 'Canegrass' was far from pristine being part of the very large Stuart Creek Station which we had been on since crossing the dog fence. Their lease extended north to beyond the old Ghan Railway, to the Oodnadatta track and Lake Eyre South, and also away to the east and west. To the left of our track on entering the area from the south was a large excavated pastoral dam and cattle yards. A fence line and station track crossed our line and the area was extensively grazed by the cattle on the property.

On entering the Canegrass area we all had travelled along the smooth natural surface of our marked line for 12 km to the northern end of the open area. The KPC had previously agreed to plot areas of concern with us but were now not prepared to do so. We had an extremely rough bumpy ride returning across country on the other side of, and well clear of the swamp where they were suggesting a possible alternative route for the road. The land to the east of the swamp sloped up gradually for about 2 km the line of very high, bare, bounding dunes, Numerous shallow gutters crossing our tracks indicated a high degree of runoff after a storm which could cause problems on a route on this eastern side of the large swamp. We were reluctant to move from the pegged line which had been fully documented in the EIS and where Government approval to construct had been received. Aboriginal groups to the north had previously declared our route as being clear of any sites and it was also supposedly beyond the recognised Kokatha tribal area.

We were to be joined at Canegrass by the RMS Manager, John Copping and Geoff Witham at mid-day and the whole party waited patiently at the south end. We waited, and waited. They had been held up and with communication difficulties it was late afternoon when we all finally met at the 40 km mark as we were heading back to Olympic Dam.

In the discussion that followed the KPC party wanted to go south and renegotiate some areas already passed by the dozer. This we refused and pursued information as to whether the KPC regarded Canegrass as a collection of sites or one large site. This was not resolved, so Mike Madigan agreed to come back to meet with the KPC and review the position after the weekend.

The RMS party then went north back to Canegrass to show the manager and legal man what they could of it in the failing light. The others headed back to the camp and on to Port Augusta and Adelaide.

It was dark as we travelled back to Olympic Dam, but fortunately we arrived there just before the canteen and mess closed. We then had a meal and left for Adelaide in the charter aircraft the boss had arrived in. The children from the van park had enthusiastically helped to lay out the kerosene flare path along the sides of the airstrip for the take off. In the meantime the road building well to the south of 'Canegrass' would continue.

It was twelve days later when the next joint visit was to take place so it was back to Olympic Dam and on to Canegrass with Colin, Mike Madigan and John Wommersley to await the arrival of the KPC party for further discussions. Colin and I had a vehicle fitted with a WMC Exploration Division two way radio and there was someone on a listening brief at the RMS office in Adelaide. We couldn't contact them on arrival at Canegrass but on returning to the top of the second dune we managed a static distorted contact.

After half an hour, at 10.30, a fleet of 4 W.D. vehicles topped the large fringing dune and descended on to the sand spread where we waited. There were about twenty in the aboriginal party with Chris Charles and tagging along in one of Blue's hire vehicles were representatives of the media. The KPC group swung their vehicles around and parked on an area we and the Government people had been very careful not to disturb. It was one of the best artifact strewn camp sites we had seen.

The group settled in and spent the rest of the morning in private conference seated in a semicircle on the sand spread at the foot of the dune. The media representatives and the rest of us were excluded and left waiting, not all that patiently, a short distance away.

We sought advice from Chris Charles as to what was intended as we understood the purpose of being there was to determine if there were matters affecting our approved road alignment.

Eventually they agreed to receive us and there followed a prolonged discussion with the group still squatting in a circle on the sand. The KPC party had been supplemented by others representing the Pitjanjatjara Aboriginals and the National Federation of Land Councils. We were clearly outnumbered but entered into discussion hoping for some resolution. The mood of the meeting was distinctly cool with us being forcefully reminded of having bulldozed a site. There were also implied threats if we did not heed the KPC requirements. They also asked if the Company would fund the group for a prolonged stay to enable them to search and assess the area and define an acceptable alignment. I replied that we could not agree to pay for this as we already had approvals from the State to construct the road on the pegged route. We would however be prepared to look at further minor deviations such as had been agreed on the line to the south.

The discussions then broke up so the aboriginal Elders could travel our line to the north and inspect the area. While we awaited their return I managed to make contact with Adelaide on the radio after trying from the top of several dunes. Hugh Morgan, who was at a meeting in the office, seemed to enjoy this different means of communication. It was new to me also, so we missed the start of several sentences for a while as we wrongly pressed the transmit switch or forgot to end with 'over'. He and John Copping were interested to hear the latest and to re-assure us that we should continue to play it by ear.

We gathered again when the Elders returned but there was no resolution or moves to agree an alignment. We, with the media, were told that they were going to establish a blockade to deny us entry and that they would camp on the site. I agreed that the construction plant would not enter the Canegrass area whilst negotiations were in progress and re-iterated we could not agree to fund their stay. To avoid a stalemate Mike Madigan agreed that the Government would help them in the form of a credit for food at the Andamooka Store. The KPC then told us that they would peg out a deviation around the area over that weekend but would not consider our request for us to join them in doing so. We were also told that they were now blockading the area and on no account could we travel through to the borefield. They had come well prepared and soon had the line to the north symbolically barricaded with a couple of stakes and mulga branches. Before we left they were busy establishing their campsites and campfires among the scattered artifacts along the foot of the sand dune where we entered the 'Canegrass' site.

That weekend the blockade, showdown and other colourful stories were highlighted in the national press and continued the next week with the experts on the evening current affairs programs on TV. They didn't mention that the camp set up and the KPC vehicles were parked on and ruining one of the better Archeological sites in the area.

On the Monday it was back to Canegrass. We had agreed to travel and assess the suitability of the alternative route well to the east that the Elders had pegged over the weekend. It started 2 km short of Canegrass and started off and continued for 17 km over high mobile dunes, watercourse crossings, slopes with storm sheet flow and gullying. We travelled all the way in low gear on one of the roughest cross country runs I had experienced. We even had to deviate from their line to try and find a way over the highest of the bare mobile dunes. The line passed close to what were to me, notable camp sites with grinding stones and artifact scatters. This route had not been through the exhaustive EIS reviews as had our chosen line and it was uncertain how the State would react to such a major change. I travelled the line a second time the next day and undertook to advise the KPC the acceptance or otherwise of the proposed deviation as soon as possible. In the meantime we would not work plant closer than 2 km to the south of Canecrass or before a fence at the north end which they had nominated as beyond their territory.

A day or two later I had estimated the high additional costs that would result from using this alternative route. As well as major engineering difficulties, there were obvious ongoing environmental problems.

We advised the KPC and the State that the deviation was totally unacceptable, so the blockade continued. At that stage our dozer was cutting through dunes 6 km back and the main construction was 16 km short of Canegrass. There was disagreement among the KPC party as to whether we could walk the plant through to the north and continue work beyond the disputed area. There being no resolution when the time came the plant travelled around the area on the network of station tracks. Bill Rymill organised with Brambles to bring in a small camp and fuel via Marree to the test bore Gab 6.

The plant re-started building the road from there to the south. The first kilometre or so was built nice and straight and much wider so it could be used as a dry weather airstrip. This gave us direct access to service the crew and for emergencies. Bluey became a regular visitor in his light aircraft flying us in as needed.

Construction proceeded apace across the open gibber with ample bore water available to build a compacted smooth road at almost a kilometre a day. Crossing the wide, dry, braided, gravelly bed of Gregory Creek was the only major obstacle along the way. In mid September the camp was moved 30 km south to the start of the long north-south sand dunes. Construction continued through the closed catchments in the open swales between the low sand ridges.

The Canegrass Blockade with all it's publicity had aroused the anti-nuclear groups who saw the August/September school holidays as an opportunity to marshall their forces to blockade the mine and village. It was a major event keeping "Roxby Downs" in the headlines and on the nightly current affairs shows. The reporters and camera crews endlessly sought confrontations and incidents. The protesters came from far and near in buses, in vans, on bicycles, and were numbered in their hundreds. There to maintain law and order were a very large contingent of police in their own encampment in the village. Much to the delight of the village children and manure starved local pot plant gardeners the mounted police brought the magnificent Police Greys.

Seventy five protesters were arrested when they tried to stop the mineworkers from entering the shaft site and went before a magistrate at Andamooka. Otherwise they were mainly of nuisance value as they desperately sought to generate publicity. There were a few irresponsibles who infiltrated the van park in the dead of night. This distressed a number of the women and children when detected right alongside their vans. However they all turned out to watch the protesters in colourful procession through the village on their way to the mine. A few camped at the edge of a dune south of the mine and toughed it out there for some weeks.

I didn't get to see much of all this, becoming involved in company activities at Stawell and Bendigo and taking some leave.

The Canegrass blockade persisted through to the end of November by which time the road construction was creeping closer from the north. The work on the Pilot Plant was also progressing and there would be a need for the water from the artesian basin early in the new year. Bill Rymill and Frank Boulton had the work needed to install and run the Desalination Plant well under way. The key to having both operating was the completion of the

Negotiations at Canegrass.

The Road Barricade during 'The Blockade' at Canegrass

Borefield Road. Flooding at Gregory Creek.

Water Tanker on the Borefield Road

Borefield Road to enable water to be transported. The Labour Party Government, having given the company approval to build the road were anxious to avoid it being held up any longer.

The Minister of Mines detailed Peter Hill to get together with the Environment Department officers, the KPC, and the company people to find a solution. Peter was now known by us as 'The Commodore' due to his role in the Naval Reserve, where he might be called on at short notice to steer a warship into Port Adelaide. He managed to get the KPC to outline another alignment,this time some distance to the west of our original track. Peter had all parties marshalled at the site by mid morning on November 30th ready to travel the proposed route.

It started off through a largish depression with dense canegrass at head height,then across country bearing to the north west and ended in a tangle of dunes in the north. We had a sandwich lunch along the line by which time the wind from the north was at gale force. It was whipping up the sand and sandblasting all in its path including our eyes trying to assess the new proposed route. We covered the ground several times before returning to Olympic Dam.

John Wommersley and Colin elected to travel north on the new road which had now almost reached Canegrass and spend the night at the chicken ranch at Gab 6. Bill Rymill had told us that a thunderstorm had caused local flooding and had scoured a deep trench across the road about 30 km south of Gab 6. We pointed this out to John and Colin and told them to take extreme care as they would have to detour around it. They left in a brand new Government Land Cruiser well before dark and must have been deep in conversation when they hit the deep washed away gutter at speed. Fortunately they were able to walk away, which they did for the 30 km to Gab 6. They arrived in the middle of the night after battling the fierce headwind for hours until it moderated when it started to rain. The vehicle was less mobile as the front axle and wheels had been pushed back under the engine so they were nearly level with the front doors.

The next day I had to tell the Commodore that we could not accept the deviation as an acceptable route so it was back to the drawing board. Peter's next move was to get all parties to Canegrass on December 14th. He announced that we would be staying there until we agreed a way through. If it meant Christmas in the sand dunes in the rising temperatures - so be it.

It now appeared that a line a short distance to the east of the central swamp was a possibility, provided we headed for it directly from where our construction had stopped, 2 km back from Canegrass. It would not add to the overall length of the road and future pipelines but would add to the cost of construction due to the cross slope drainage. We started to mark it with the KPC elders. Half way along they said we had come to the end of their territory and from there we could go where it suited us to join up with the road north. We agreed to accept this line much to the joy of all concerned.

Bill Rymill and Frank Boulton soon had a grader on site to lightly blade and mark the adopted route. The section set out with the elders was in a series of long straights but John and Colin took it upon themselves to move some pegs to miss a couple of minor stone outcrops.

Chapter 23

THE DELUGE IN THE DESERT

The blockade now lifted, the road builders took a short well earned break. All involved headed home to enjoy their Christmas. Frank Boulton and Paul Johnson from Brambles had the dozer, scrapers and graders ready to roll again in the first week of January. They needed to complete the now deserted section through Canegrass promptly to allow water carting to start by mid February. Otherwise the commissioning of the desalination plant and start up of the pilot plant would be delayed.

Starting 2 km south of Canegrass immediately after New Year they were in loose mobile dunes for a start before heading north on the open expanses to the east of the low lying swamp. All was going well until Friday 7th when steady rain and muddy conditions stopped the plant from operating. The rain set in and the new road also became wet and slippery. It took 8 hours and the assistance of John McKinnon in another vehicle to rescue Frank before slithering back the 45 km to Olympic Dam. Behind them were deep twisted wheel ruts on all but the short sections in sand or with a gravel base.

The rain continued into the weekend with muddy conditions also stopping work on the pilot plant. On the Sunday Frank arranged with Bluey to fly him at low level along the road to the borefield. There was evidence of some damage at short washed out sections south of canegrass. Further north it was clear that the rainfalls had been much heavier with water ponded over the road in many of the hollows between the dunes. As they neared Gregory Creek there was water everywhere with the normally dry creek bed carrying a major flow. The road crossing was flooded with a fast flowing torrent of uncertain depth for a width of about 150 metres. The road on to Gab 6 seemed to be intact but it was obviously too soft to land the aircraft on the widened section of road. Gregory Creek as it passed by the bore was also flowing deep and fast as it sped on its way to Lake Eyre South. The area around the bore and the built up road turnaround were above the flood and unharmed. Frank called Bill Rymill on landing saying it was "wetter than the west coast of Tasmania" and asked that he bring a video camera when he came up on the Monday.

It was obvious that there would be major repairs and rebuilding of completed sections of road in addition to the completion of construction through the canegrass area. On Monday Bill, Frank and I used the charter aircraft from Adelaide to look at the situation. The pilot, nicknamed 'Handlebars' was a regular on the run to Olympic Dam and with his red hair and fighter pilot moustache flew accordingly. He was delighted when we asked him to fly low along the length of the road to the borefield. This was both to let Bill use a video camera to record the very unusual affects and for us to get a feel for the likely cost and time to get the road trafficable. Fly low he did, That plus some tight circuits to get a second look here and there made it a very memorable ride. It also made 'Handlebars' day.

The dry white salt crust of Lake Eyre South was now covered with the incoming floodwaters. The total rainfall as measured at Gab 6 was more than 300 millimetres in what is the driest part of the continent. The heaviest falls were fairly localised in the catchments feeding the Lake from the northwest and from the south. The falls tapered away to the south with less than 75 mm at Olympic Dam and almost none at Woomera. Nevertheless it was a sharp reminder that heavy rainfalls of similar magnitude were possible in the project area even though records to the time indicated otherwise.

Frank and Bill set out to muster all the plant available to get the road completion and repairs under way quickly as the country dried out in the fierce mid summer heat. Bluey again took to the air to help out, landing on the Oodnadatta Track from where Frank walked in to Gab 6 to check if the road was dry enough to land there. Fortunately a vehicle that had been stranded there was waiting for him. This let him check for damage to the road south as far as the still flooded and washed out Gregory Creek crossing.   The rush to get the road serviceable for the water tankers to start filling the small storage at the now completed desalination plant was now on. Working long hours for seven days a week resulted in the first Top Transport tanker being able to load up at Gab 6 and make the first trip in mid Febuary. As the new road compacted they added a trailer with a second tank and later travelled in triple configuration carting 60 tonnes a trip.

To ensure the new road, now known widely as the Borefield Road started off with some landmarks we provided kilometre marks every 5 km and two signs. One proclaimed "Wommersley's Wobble" and the other where the new vehicle came to grief as "Woolard's Washaway." The road transport of the raw artesian basin water would continue until a pipeline was installed if and when the joint venturers committed to a major mining project. The free flow from the bore at Gab 6 now fed through a large tank so as not to affect the long term test. The tank held enough to fully load the road train and refilled quickly.

The 400 kilolitres a day desalination plant was now successfully commissioned. The reject brine from the plant was shandied into the raw water pumped to the pilot metallurgical plant. The mixture became known as "process water." The clear potable desalinated water was pumped to the mine, pilot plant and to the large storage tank in the village. We were no longer dependent on the supply from Woomera.

It did not take long for the Borefield Road to attract a multitude of users. The surface was very good other than in the wet. It was intersected by a station track about 5 km short of Gab 6 and this gave access to the Oodnadatta Track at Bopeechee, then on to Marree or William Creek following the old Ghan railway. The heavy rains that had disrupted our road construction and the resulting major stream flows had filled Lake Eyre South sufficiently to overflow through the Goyder Channel and into the much larger Lake Eyre North. This was the first time this had been recorded with previous fillings being overflow from the north to south. The water in the south lake was readily visible from the Oodnadatta Track and visiting via the Borefield Road became a popular weekend excursion for the Olympic Dam residents.

The new road also gave better access for those enthusiasts, mostly from far away Adelaide, who grouped together to restore and preserve a section of the old Ghan railway. They were few in numbers but headed by Simon Coxon started to make regular visits, basing themselves at the disused but sound Curdimurka station buildings. Calling themselves "The Ghan Railway Preservation Society" they planned to retain and maintain several miles of old track. This included the historical multi-span steel bridge to the west of the station over Stuart Creek. The artesian bore at the station was still flowing nicely and gave an assured supply of water. The Olympic Dam people were supportive and helped where possible with supply of transport and materials. A year or two later an unusual fund raising "Black Tie Dinner" was held in the outback wilderness and isolation at the old railway station buildings. The idea of a Saturday night in such unusual circumstances attracted guests from near and far. All camped out in the desert after a fairly ordinary but much appreciated meal prepared in extremely difficult conditions. In later years the instigators decided that an "Outback Ball" would be more appropriate. This became a well publicised and popular bi-annual event lighting up the desert with the largest crowds ever to assemble at Curdimurka.

The pastoralists also found the new road a more direct way to truck their cattle in and out to market. It was in fact a private road but it's attraction to commercial and recreational users was recognized and use permitted. The only concerns were for safety and to prevent damage by use in wet weather. The road tankers were on the road night and day but visible from the distance other than at dune crossings. Not all that long after it was completed the route started to appear on road maps and in articles on outback travel.

Chapter 24

THE PILOT PLANT AND PROJECT PLANNING

The pilot plant started operating in March of 1984 adding to the on site workforce. The 12 houses, 50 caravan sites and single quarters for 250 were fully occupied. The airstrip formation had been widened to 90 metres and it was now licensed. There was also a growing number of privately occupied sites in the village industrial area. The population welcomed the very high quality water being produced from the smoothly operating desalination plant. The only complaint was from the visiting ladies' hairdresser who could not get the hair to set with the very pure water. This was soon fixed by adjusting the plant so it wasn't quite so efficient.

There was now more confidence that a major project would follow the findings from the Pilot Plant metallurgical process testing. The mine workings were now extending well out from the shaft permitting the directional drilling for core samples from underground. This enabled the speedier, cheaper and more precise delineation of potential ore zones than from closely spaced holes drilled from the surface. In late 1983 the Joint Venturers announced a "Probable Ore Reserve" of 450 million tonnes.

All this added to the sense of optimism after the passing of the years without any certainty of there being a mining project. It had been frustrating and unsettling for both those at Olympic Dam and those in the RMS Adelaide office. Being attached to the Perth Engineering Group I had been effected to a much lesser extent as I became increasingly involved in other areas.

They were getting to know me by my first name at the airport having flown out over 120 times in the four years to the end of 1983 with only about half of these being to Olympic Dam. Although most of the away trips were to Perth, I was becoming a regular visitor to the WMC gold mine at Stawell in Victoria. This started when problems arose after a large underestimation of the amount of water to be pumped out to dewater old mine workings.

I called in Cec Forbes, our groundwater consultant who quickly produced a realistic estimate of the volume to be removed to allow mining to proceed. Then followed the problem of disposing of the large amounts of moderately saline water from the pumps. The solutions proposed by the consultants that had been dealing with these matters were unacceptable and we elected to buy land and use extensive shallow evaporation ponds. The reviews of the progress of the dewatering and other work meant a return to Stawell each month. Brian Micke, the Mine Manager had particularly asked that I be there in late January of 1984. The councillors from a Municipality at Bendigo were to visit. Their purpose was to see what impact the mining operation had on the town as WMC were starting exploration on the old Bendigo goldfield. I looked forward to that day after learning that Hugh Mason, their Engineer would be with the council visitors. Hughie was the Engineer at Yackandandah Shire when I was at the nearby Upper Murray Shire. We were both 1926 models, had been close friends, but had lost touch over the years.

There was a second reason for wanting to catch up with Hughie. I felt he just might be at the stage of looking for something completely different. It was time that RMS had a mature and experienced engineer in the office who could concentrate full time on all aspects of the proposed township development. Hugh certainly fitted the bill as regards municipal experience with a record of sound engineering and the ability to get on with people. The visitors duly arrived and I was delegated to show Hughie around for the day. I was delighted to learn that he was indeed thinking of leaving the Council. He was considering a consultancy but showed interest when I mentioned the possibility of him joining RMS to look after the town work. I then had to convince John Copping that there was more to the planning and liaison with the State authorities than I could reasonably handle.

There was now increasing emphasis on determining precise information as to the layout, zoning, design criteria and policies for the future township. John Harris and Peter Hill were being pressured for details by the State Departments who were to provide schooling, medical facilities and other services. We held town meetings monthly and had Brian McKay busy producing plans with control designs. The proposed roads were starting to get names starting with Axehead Road which closely followed the old station track. This road provided the initial access to the townsite from the Woomera Road and basically controlled the rest of the development. It was the principal drainage line from the higher ground in the east to the low lying storm pondage areas to the west of the habitable subdivisions. The very infrequent heavy rainfalls meant we could rely on the properly designed roads to carry storm flows out of the town. This would eliminate the need for stormwater pipelines throughout the township other than for culverting under the road at the western boundary. Beyond this the storm runoff could be collected and added to reclaimed sewer effluent for watering ovals and other grassed areas. The general low lying area would be set aside for a future golf course where short term ponding in the hollows could be tolerated.

I was relieved when Hughie advised his interest in joining us and got approval for him to come to Adelaide and check us out. In late Febuary Hugh and I flew to Olympic Dam , had a quick look around the village and mine, then down to look at the townsite. We turned off the Woomera Road along the Axehead track then south to the top of a high dune overlooking the future town centre and commercial area. It was an expanse of red sand, with dotted mulgas and saltbush. Beyond on the lower ground were myalls, more mulga and saltbush, then more dunes with stands of native pines.

Hughie was overcome with the emptiness and the obvious challenge of changing it into a large, people friendly, and welcoming township in the desert.

Years later I learned that the poet in him remembered :-

"There was stillness in the sand dunes
When first I trod the ground
It was a strange and eerie place
I wondered what I'd found."

Hugh Mason.

The opportunity to be closely involved in creating a brand new town in the harsh arid outback would be a new experience for Hughie and he indicated his interest. It was not until May that we were able to make a final offer to him which he accepted.

By then the decision had been taken to have a major engineering contractor carry out a Technical Study for the development of the Olympic Dam Project. In the office we had considered four options in a detailed 'Pre-feasibility Study.' These were for alternative staged developments leading to the output of 150,000 tonnes of copper a year. We also prepared a project description to assist in interviews with various contractors competing to carry out the independent Technical Study. In July Fluor Australia Pty. Limited were appointed. Shortly after Malcolm (Mal) Hayes from Fluor and a number of their technical and specialist staff and consultants moved into, or visited the office to get the study under way. They were to review the studies, test work, findings, planning and site installation carried out to date, then produce a report and cost estimates for the works required to get the production started. Not long after Fluor had started, RMS had decided that initially the project should be developed at a smaller rate of mine production. This was to be based on using high grade ore to produce 2000 tonnes per year of uranium oxide plus the contained copper, gold and silver.

Hughie joined us at the end of June on a contract basis and soon was catching up on the present state of play in respect to the town. There were a number of other changes in the office. John Copping left returning to W.A. and Tony Palmer moved in as Manager, Eastern Operations which included RMS and Olympic Dam, the Stawell mine, and the Kingston coal investigations. The work at the old Bendigo Goldfields also came under his control.

George and Gillian White had left Olympic Dam and were now at Bendigo in charge of the efforts of looking for more gold. I also ended up as a regular visitor when the problems of dewatering old workings and safe disposal of the arsenic rich water arose. Again Cec Forbes featured in assessing the quantities to be pumped. Bill Rymill and family moved up to Olympic Dam where Bill replaced Gerry Gould as Chief Engineer looking after area maintenance including the now busy pilot plant and mine. There was also the continuing care of the 200 km of busy unsealed roads from Woomera to the Borefield.

In the office we soon got used to having the Fluor people at our elbow. Mal Hayes managed their efforts and with his friendly, easy going manner was well accepted by the RMS staff he was now dealing with on a daily basis. There were endless meetings bringing his specialists up to date with what we knew, didn't know, and what we thought we knew.

John Harris was pleased to have Hughie available to concentrate soley on the town and village. There had been a constant demand for industrial sites in the village for equipment suppliers and maintenance contractors servicing mine, pilot plant and infrastructure. We had surveyed allotments and progressively extended roads and services. Peter Hill required us to assess the costs of each of the allotments. This was so the occupants who were making substantial improvements could be made aware of the prices for purchase at a later date, if and when notice of Project Development was given to the State. Peter was aware that I had costed all allotments at both Kambalda and at Laverton in W.A. so was insistent that I be involved. The indenture was specific in that only the direct costs in developing the lots could be applied to their value and sales. Eventually we arrived at satisfactory answers.

The only vehicle fuel available on site at the time was at the drill workshop and store near the core farm which was causing congestion and problems. The future through road from the town to the mine would be to the east of the camp and van park to be well clear of the airstrip. We decided to build the short section of this road needed to join in to the existing run to the mine and selected a site for a service station along it. BP were given the right to develop the facility which they promptly did. Bluey and Jan Lavrick became the proprietors and added a vehicle hire and travel service to their activities. They were also kept busy servicing the regular air services and refuelling at the now busy airport. The convenience of stepping from an aircraft, into a hired 4WD Land Cruiser and be on the way to the borefield in a matter of minutes was much appreciated. I was making regular trips to keep up with the continued monitoring and test drilling. The Borefield Road was maintained in such good shape that it took little over an hour to reach Gab 6. It was a busy spot with the water tankers arriving to load on a regular basis. The 'chicken ranch' was usually hosting some sort of 'oligists' on continuing studies of the mound springs. The monitoring of drawdowns from flow tests and from all the test drilling were giving us detailed information on the Borefied A area. It was the most detailed study of the nature and behavior of the aquifers in that region.

AGC were now able to develop computer models to confidently predict the affects of drawing from the bores at various rates. It became apparent that Borefield A could produce more than the initial estimates without exceeding the drawdown limitations. The full details of all the monitoring, drilling, modelling and findings were reported to the Mines and Water Supply Departments on a continuous basis.

Peter Hill and Geoff Whitham decided that we should make a goodwill visit to Marree. As well as assuring the locals that our activities would in no way deplete their bore water supply it would let them question us on any concerns they might have. Five of us made the trip staying overnight at the hotel where they looked after us very well. It was a typical outback two storey pub with all of the bedrooms upstairs. They must have thought WMC were a bit short of travel funds so three of us were settled in one bedroom and two in another. It didn't trouble us nor did the excellent dinner which was a choice of steak or steak.

The meeting was well attended by town people and the pastoralists from near and far. It was over and done with in quick time and as outback custom would have it we withdrew to the bar. It was not to be an early night. Geoff Witham maintains to this day that before the party broke up I had promised a set of starting gates for the Marree race track. I very distinctly remember that it was Geoff who made the promise. In any case we got to know and establish good relations with some of the characters of the region. The two of us that shared with Peter Hill were fortunate.(?) In true navy fashion he brought us a cup of tea in bed at 5am.

When Hughie had settled in we took the opportunity to visit the WMC operations in Western Australia. This was to let him see the standards of housing, services and facilities at the predominantly company towns of Kambalda, Laverton and Leinster. The housing officers at each place were full of information as to what policies and housing details were good or bad and had excellent advice to offer. Kambalda had been developed in a period of full employment and rentals had been set at a low weekly rate to help attract labour. House and land cost at the time were in the order of $14,000 for a three bedroom home. They were offered for sale to employees on favourable terms at cost but there were very few takers because of the low rentals. The company as the landlord was responsible for the ongoing excessive costs of maintenance and repairs, as well as rates, and insurances. All of this together with a collection of typical house plans were indications of the policy matters that needed to be determined in the immediate future.

Before leaving Perth we visited the WMC office so Hugh could meet Peter Webster and see the workings of the engineering group. Time was also spent with Max Hannel who had featured in the architectural input for a number of major WMC buildings for us in the past. Max, in his first years after graduation, had been signed on by Poseidon as my offsider for the Laverton township development. He then started his own architectural practice as a sole practitioner. He seemed to be able to produce what mining industry clients wanted with a minimum of fuss and within a tight budget.

Hughie had quickly settled into the office and soon got to know those he would be dealing with and what stage the planning for developing the town had reached. At our regular meetings with John Harris he continued to press for information so the planning by the State for the schooling, public buildings, and government housing could be advanced. It was now realised that project development was probable and that we would would be quick off the mark with the town construction once the decision was made. The State Public Building Architects wanted to be ready with their plans so that the school and other community facilities could be constructed promptly. They needed confirmation of the site areas, details of levels, services locations, and much more. All parties were keen to ensure that the architectural style of the public buildings was compatible with that adopted by the company. (Or vice-versa.) Fortunately as far as we were concerned, and thanks to Geoff Witham's input to the Indenture, we only had to deal with the Mines Department and not with each individual Government Department. Hughie and our town consultants, Ulman and Nolan, were still to work closely with the various departments but with John Harris or Peter Hill there to resolve any difficulties.

We heeded the requests and set out to find architects to prepare sketch concept presentations for the style of development for the town centre. A number of Adelaide architectural firms had been persistently offering their services over the years. We decided to conduct interviews, give them the opportunity to put their ideas, and sort out who might be best suited for ongoing work. Hughie and I got frustrated very quickly. They found it hard to appreciate that we needed them to show by way of illustrated sketches and brief descriptions how they envisaged the streetscape might look. This was to give them the chance to show their talents with the promise of lots of work if they were appointed to handle our work. What we wanted was something that the company and the Public Buildings officers could accept as setting the scene or a theme for the town.

The primary consideration of those we interviewed seemed to be what we were prepared to pay to have the presentation prepared and how long it was going to take. Finally in desperation we rang Max Hannel in Perth. Two or three days later nicely coloured vistas of the what the town centre might look like arrived in the overnight mail bag from Perth office. Max and his artistic draftsman had wasted no time and the result was just what we were after. It was the first glimpse of what a part of the new town in the desert dunes might look like. It was something to be displayed, considered, modified and taken the further steps to reality. At the time, we weren't able to extend to Max the probability of further work so we asked for an account which eventually came in at well under $1000.

Although the future town would only be a very small part of the overall project development it was the part that would be the focus of much attention. The massive underground mining works, and the intricate web of the multi-mineral metallurgical plants would be largely hidden from the public eye. They were to be the whole reason for being there but it would be the town and its facilities that everyone could relate to. The simple sketches were used by the Joint Venture partners to illustrate that this was to be a town of substance. The sketches even made it into the daily newspapers where any descriptive information on the new town in the desert was eagerly devoured.

The State Highways Department were also anxious to be up to date with their planning for the sealed roads they had to build at our cost. These were to be from Stuart Highway at Pimba to the Mine Site boundary with connections to the town and the airstrip. They had already negotiated a temporary route through Woomera when the town was opened up and the control gates removed. This shortened the distance by eliminating the rough circuitous detour around the town to get to Phillip Ponds. However the traffic noise from the increasing numbers of heavy trucks that were finding their way through the centre of Woomera was not well received by the diminishing population. The traffic was then re-directed along perimeter roads, well clear of most of the residential areas and the outcry lessened.

To make sure the State road planners were up to date and to iron out details of the road to be built, Peter Hill kept arranging meetings at the Highways Department office. I was the only Company person with road design and building experience so I was taken along each time. The meetings were held in the Board room with some of the Highways Commissioners in attendance. This was probably not because of the importance of the discussions, but that Olympic Dam was a hot and topical subject. It gave them an opportunity to get some scanty information direct rather than from the suspect interpretations from sections of the media.

The road planners decided on a resurvey to select the route from Woomera to the north across the gibber and beyond, using a helicopter and the new fangled 'Global Positioning' system technology. This was completed quickly, (at great cost) and led them to the conclusion to follow the alignment we had adopted when we built our existing road. They also had new environmental studies carried out along the route.

The increasing activity on site also lured the anti-nuclear activists to mount another visit to site and demonstration in the August-September school holidays. Once again the Police Greys and the media reporters had a trip to the outback, the city policing diminished and the irritating invasion to site was repeated. There were attempts to prevent the miners entering the shaft site which were thwarted when they used a back gate. The Pilot Plant kept operating treating 5 tonnes of ore an hour and the visiting police enjoyed the sunshine. The neighbouring pastoralists gave unqualified support to the village residents and the protesters were made decidedly unwelcome at Andamooka. Eventually the school holidays were over and peace was restored at the mine.

There was now a lot of involvement with Mal Hayes and his specialist staff. The detailed technical study which they were progressing would hopefully lead to the decision to proceed with the project development. They reviewed the physical, planning and testing work that had been undertaken and was proposed by RMS. There were assessments of all the legal,technical and financial aspects of developing and maintaining a major mine. They checked everything from the geological ore reserves, mining methods, metallurgical processes, proposals for tailings storage to all aspects of our engineering proposals. They regarded water supplies as critical and had hydro-geologists check all our assessments of continuing usage from the artesian basin. The provisions of the indenture, particularly as they related to water supplies and the town were considered as were environmental restraints. They checked that our estimated numbers of employees and staff for each year of pre-production and operating were reasonable. Then that our allowances for all items for the construction workforce accommodation, township and housing were realistic.

By the end of 1984 the initial pilot plant program was complete and the basic metallurgical processes confirmed. There were also 7 km of underground drives to verify ore blocks, grades and the competence of the rock for large scale mining. On the surface we had finalised an overall land use plan detailing the boundaries of the Special Mining Lease, the Municipality and the Town. All this was taken into account by the Flour people and when all was accepted or revised they prepared detailed cost estimates and provisions and documented their findings. This detailed 'Technical Study', in bound volumes was presented to the Joint Venturers for their consideration in March 1985. They in turn then carried out a 'Final Feasibility Study' in the following month. By the end of June the confidence among the RMS staff that the project would be given the go ahead in the not too distant future was at an all time high. The meetings with the State authorities increased with more detailed planning and discussions.

Telecom also set up a group from their multiplicity of departments and specialities to meet with us and be advised what our communication needs would be. Headed by John Grivell they set up monthly meetings to plan the works required to fully service the mine and new town. I always tried to take someone with me to the meetings or I would be a lone voice among the eight to ten Telecom people at the large table. Glen Weir, an electrical engineer recently transferred in from Kambalda was able to fill the gaps on the needs for underground phones. Hughie soon found a site for the Telephone Exchange in the town. There was to be a state of the art incoming optical fibre cable from Woomera which would link to the cable being installed from Darwin to Adelaide. A route for ploughing in this cable following but not conflicting with the future road construction was agreed as was entry across the dunes to the town exchange.

Hughie and John Harris determined where all services would be located in the town streets and fitted in a place for the telecom cables. We had already installed cables throughout the village camp and Industrial area and up to the mine, pilot and desalination plant. The company would have it's own permanent private exchange, (PABX in telecom parlance) at the future Administration Office located in a 'Central Services Area' near the Metallurgical Plant. A 200 pair cable would link this to the Telecom Exchange in the town.

A communications tower would be sited on a dune just to the north of the town. This would enable temporary additional radio link incoming lines to be provided during the busy construction period. It would also carry T.V. transmitters taking over from the temporary set up at the village and 2 way radio aerials for the company mobile communications installations. The tower site was listed as a high priority for power supply. This was all terrific except that the company would have to pay the costs of over a couple of million dollars.

The surge of confidence meant that we could get final plans prepared and adopted for the six hundred and fifty houses, the single quarters complex and the caravan park in the town. John Harris and Peter Hill ensured that the State Public Buildings Department were also in readiness for a start. They completed the plans for the educational facilities, the community centre, municipal offices and health centre. Those for the swimming centre, courts and playing fields were to follow. The company agreed to provide a Licensed Club as an early development. It could be some time before a tavern or other licensed premises were built by some private developer. Max Hannel was commissioned to prepare drawings and a specification for the club which he did in quick time.

Bob Crew was totally in control at Olympic Dam and with the mine development going at a cracking pace was keen to finalize details of the main office and stores buildings. Hughie and I quickly agreed that we could kill two birds and make it easy for ourselves. Bob wanted to be involved in the layout planning of the office. Also he and his wife, Lyn needed to have input into the planning of the Manager's house. Bob was fully committed during the week but would make himself available for a weekend. We arranged for Max Hannel to come to Adelaide, fly up to site on the Friday and to spend the weekend with Bob and Lyn planning these buildings. Fortunately they all spoke the same language and when Max returned to Adelaide on the Monday all was agreed.

The planning for the mine, metallurgical plant tailings storages and services was also proceeding apace. Mal Hayes and a number of his staff had continued on in the office after the completion of their study. With Mike Softley they were finalising the plant location and layout. I became involved firstly in respect to the drainage, design of the roads, and the provision of water supplies to the plant, the adjoining main office and stores area. Because of the uranium in the ore the whole of the mine surrounds, ore stockpiles and metallurgical plant were designated as "dirty" areas. This meant access would be strictly controlled, and the whole area surrounded with security fencing. The storm drainage and accidental spills within the dirty area had to be contained within the fenced boundaries. Entry into the plant and mine sites would only be at manned gatehouses on the security fenceline. All personnel working inside the dirty area would be required to change into work clothes at the start of their shift. On leaving they would shower leaving the working gear to be washed at laundries attached to the change rooms.

Outside the fence adjacent to the plant in the "clean" area there would be the main administration office and warehouse and storeyard, general workshop, and large bus and car parks.

At the plant gatehouse there was also to be a heavy duty weighbridge and an automatic wheel wash for outgoing vehicles. The entry control at the Whenan Shaft was already in place as were small changerooms but substantial additions would be required. The security fencing enclosing the dirty area would need to be carefully sited so all storm drainage was inwards.

Both the shaft and the soon to be completed decline road access to the mine workings were in the same line of swale as the proposed metallurgical plant. The swale was bounded to the north and south by long dunes. The two sites were however in different closed catchments with a low cross ridge separating them. The stormwater runoff from the large mine catchment all drained towards the now filled claypan where the shaft was sunk. Bob Crew had initially scoffed at the excavations that I insisted be provided to catch the stormwater if and when the then current drought ended. It was not long after that a 70mm rainfall filled the stormwater ponds to the brim. Fortunately the mine remained high and dry and Bob was on the phone suggesting that we dig some more ponds.

It was critical that the drainage and disposal of storm runoff from the extensive roofed and paved areas of the metallurgical plant be carefully considered. I was aghast when first sighting the layout plan to see that the plant occupied the centre of the open swale at the low point of the catchment. In that position it also meant that the boundary fence of the "clean" area to the south was in a tangle of high dunes half way up the slope. It was soon agreed that the whole metallurgical plant layout be moved 100 metres or so to the south on to the wide elevated sand ridges. This meant that the site preparation earthworks could be designed so that the southern boundary fence was on the catchment boundary between the clean and dirty areas.

This put the plant at a higher elevation so it could be readily drained along depressed roadways. A large stormwater collection pond could then be excavated at the low point of the swale to the north west of the plant. This would ensure that adjoining roads, switchyard and service corridors would not be flooded. Two smaller ponds would be needed at the west end of the plant. Between them all they would provide for the drainage of the total plant catchment. The runoff from the clean area all naturally flowed to a large claypan bounding it to the south.

There was a great deal of earthmoving to take place to ready the total area for the major construction. Kinhill were given the task of preparing plans and specification for the site preparation and provision of roads and services. To make doubly sure of satisfactory drainage I sat at a drawing board and set out the road and earthworks levels in detail. I had ensured that nearly all stormflow would be carried as surface drainage in the side channels of the depressed roadways to quickly discharge into the stormwater ponds. This eliminated underground pipe drains other than under the roads adjacent to the ponds and meant any plant spills and silting could be readily and quickly cleaned up. It did however increase the quantities of cut and fill required for the site earthworks. This roughly drawn detail was passed on to the consultants to incorporate into the siteworks construction plans. When these landed on my desk the levels, grades and drainage lines were completely different and did not provide the falls that were needed. It was back to the drawing board as my original plan had somehow got lost. Second time around the final earthworks drawings were in order, and we could proceed to add road and services detail in conjunction with Mike Softley and his plan designers.

Chapter 25

PROJECT DEVELOPMENT

S ir Arvi Parbo and his B.P equivalent finally announced the 'Go Ahead' on December 8th of 1985.

The Stage 1 works would provide a smaller project than covered by the EIS. It would yield 48,000 tonnes of copper, 1900 tonnes of uranium oxide, 27,000 ounces of gold and over half a million ounces of silver per year. The cost would be about six hundred million dollars and the town would initially provide for a population of about two thousand. The news was received with elation both on site and in the Adelaide office.

The Company personnel to manage the development were then appointed by the Director of Operations, Keith Parry.

Doug Marshall, a Mechanical Engineer with years of experience with WMC in the Kalgoorlie area moved to Adelaide in supreme command as 'Project Manager.'

Mike Softley, as 'Engineering Manager' was to oversee the design and construction of the Metallurgical Plant and Mine installations. A group of engineers from Perth moved over to assist him and included the return of Jeremy Folwell to Adelaide. Fluor personnel with Mal Hayes would be doing the design and onsite contract supervision under the watchful eyes of Mike and his offsiders.

Hughie was appointed 'Project Engineer - Township' to look after all aspects of the town development and works at the Olympic Dam village. Kinhill would be work under his direction on the detailed design and tendering. Hugh insisted that Brian McKay from Ulman and Nolan be retained as the onsite manager in charge of the town construction.

I became 'Project Engineer - Infrastructure' which covered a wide miscellaneous range of activities. Included were all aspects of water supplies ranging from the borefield through to the desalination plant, storages and supply to all areas. There was also the power line from Woomera, tailings storages, site preparation at the mine and plant, roads, offices, stores, change rooms and miles of fencing, Kinhill were to attend to the design and the construction supervision. They soon had a number of their design staff in our office supplemented by contracts officers and cost controllers.

To ensure everything ran smoothly we were joined by David Manning as Administration Manager who had been in that position at Kambalda since it's first years of operation. To house the large influx of Fluor designers for the

Metallurgical Plant and Mine installations another large office in Greenhill Road was leased. Mike and his assisting engineers occupied the same building being responsible for review and approval of the design detail. They also ensured that where possible there was standardisation of plant and components in conformity with other WMC operations.

Unfortunately not long after the appointments were made and work was getting under way we learned of the death of Keith Parry in Perth. It was a very sad loss as Keith had been an ardent and very capable supporter of the project as well as a close friend.

There would be massive amounts of concrete and crushed stone products needed for use by the contractors in all areas. To ensure the availability at prices we could quote in our tender documents we called for proposals from companies active in this field. The small quarry to the north of the mine would no longer be accessible so an alternative would have to be used. Some years previously I had noticed what appeared to be a significant outcrop of limestone along the Axehead track a few kilometres to the west of the townsite. We had it tested, and found it to be a dolomitic limestone suitable for concrete and road construction.

We had then taken out an extractive minerals license and were now able to make it available to the chosen operator for quarrying and production of crushed stone products and concrete. We eventually chose the Readymix Group who already had a concrete plant in the village to sub-lease and open a quarry and agree to supply products at nominated rates. A bore into the underlying aquifer provided a saline water supply for dust suppression at the quarry and crushing plant. It was also sufficient to fill the needs for the town construction.

To provide an early supply of potable water at the townsite we installed some small precast concrete tanks. The construction at the town of a 10 megalitre earthen storage with plastic lining and floating cover was given priority. Pending the installation of a pipeline from the desalination plant water was trucked in from there an from Woomera.

There was also some urgency to get the contract for the earthworks, roads, and services at the metallurgical plant site and adjacent clean area under way. This involved deep cuts and fills on the dunes and thorough compaction to take the heavy loadings that were to follow. Jack Clarke, the long time resident drilling supervisor had drilled some more salt water bores for use in the earthworks construction.

As soon as a road into the plant site was trafficable, Atco units were dragged in as construction offices and occupied by the Fluor and Kinhill contract supervisory staff. Contracts were let and the work got under way. Fortunately the excavated sand from the tops of the dunes compacted solidly in the fills to form a solid base for the plant foundations.

Readymix had wasted no time in establishing a quarry and crushing plant at the limestone outcrop to the west of the town. Crushed stone road base material was soon flowing into the metallurgical plant area to establish all weather access to all points. Hughie also had contractors hard at it establishing the road network in the town and building the direct town to mine road connection. This he very aptly named Olympic Way.

The State also got into the naming game with the Cabinet after due consideration issuing a decree that the new town was to be called 'ROXBY DOWNS.' One discarded alternative with some local support was 'AXEHEAD.' Doug not to be outdone decided that the rectangular road network in the metallurgical plant would be named numerically with streets in one direction and avenues in the other. There was no argument.

Hugh and Doug were also giving priority to establishing the additional single quarters to accommodate the large construction workforce that would be soon be on site. Additional bunkhouses were landed at the original Village camp raising the capacity to 400. The kitchen, mess and wet canteen were extended and further caravan park sites provided. This area now became 'Camp 1.' Another self contained camp with 500 beds was assembled in a swale a kilometre to the south and became 'Camp 2.'

Life in the Adelaide offices was also hectic. The influx of Kinhill design and contract staff in the RMS office were turning out masses of plans and paperwork for Hughie and I to review, check and approve. Fluor had built up an even larger workforce in the other office. Here under the oversight of Mike and his team, with input from Henry, they were coming to grips with the complexities of designing and contracting the four plants for the metallurgical processes. They also were covering the mine underground and surface facilities including ore handling to and recovery from surface stockpiles.

Doug, Mike, Hughie and I were now flying to site at 6.30 a.m. every Wednesday morning, staying overnight and returning late on the Thursday afternoon. Most times Mal Hayes from Fluor and Winston Jackson from Kinhill would also be along. On arrival we would assemble with the site supervisors for a weekly project meeting and later catch up with the works in progress or about to start. It was usually well after six before we made a quick visit to the canteen followed by a late dinner in the mess.

Doug had ensured that we had permanent accommodation set aside for us in the camp as befitted our exalted positions. These were two bedroom bunkhouses with a small lounge and bathroom. We could safely leave our gear, plans and personal items in them from week to week and they were in a quiet location in the camp. Hugh and Winston shared one of the units which they managed to furnish quite comfortably. It immediately became a meeting place for all at the end of the very long day. There, usually with a stubby or Scotch and ice, we could let off steam and cure the ills of the world until thrown out. It was a real saver, enabling us to establish a rapport with each other and to receive a bit of undetectable moral support when there was a need.

I was giving priority to the drilling of production bores at borefield A and the pipeline system to deliver the water to Olympic Dam. The survey along the 110 km pipeline route showed a rising grade for the first 30 to 40 km. We intended to pump the flow the full 110 km from a single pumping station at a collector pond at Gab 6. That meant we would need pipes capable of withstanding very high pressures for at least half of the length. Tubemakers, who were the manufacturers of ductile iron pipes were anxious to show us pipelaying in progress at a site in New South Wales. Charles Duncan from Kinhill and I took the early flight to Melbourne, then on to Wagga in NSW. Tubemakers had a car waiting to take us on to somewhere beyond Cootamundra. There, contractor Max Wallace had a team busy installing a ductile iron pipeline in a most efficient manner. The trenching, pipe preparation and laying, backfilling and tidying up was carried out speedily and effectively. We were particularly interested in how a continuous protective polythene sleeve was fitted and joined over the pipes. This was to protect the iron from later corrosion from saline ground conditions.

So great was my interest that I casually leaned on one of the stationary tractors to study the activity. Without any warning something with the force of a brick projectile hit me dangerously high on my inside thigh. As I returned to ground level I noticed the tractor owner's blue heeler running off. Inspection revealed he had drawn lots of blood without even tearing the pants. The contractor and Tubemakers people were most apologetic but I had to admit the dog was only protecting his master's tractor. We called in at the ambulance station on the way back through Temora. The officer there kindly applied some antiseptic which he said might hurt a bit. When they picked me up off the floor we drove on to Wagga where bruised, bloodied and sore I caught the plane back to Melbourne and on to Adelaide.

That Friday after work when we gathered in the office for a few drinks as was our custom, Mike had prepared a surprise to recognise my misfortune. Presented formally before the assembly was a 'Blue Heeler Survival Kit.'

The Kit consisted of a tin of "Pal" dog food, a can opener, a dish, a band aid and the following typed set of instructions:-

KAY NINE AGRESSION SUPRESSION CORPORATION

BLUE HEELER SURVIVAL KIT

INSTRUCTIONS

(1)   If you see a hungry Blue Heeler approaching reach for can of "Pal", open quickly with can opener provided, empty contents of can into bowl provided and place bowl in front of the Blue Heeler. Note - proceed with alacrity.

(2)   If the above procedure is unsuccessful and the Blue Heeler's approaches cannot be resisted we wish you luck. If necessary use the band aid which is provided.

(3)   Never lash out and attempt to kick a Blue Heeler after it has bitten you. The breed has evolved over many years of biting cows heels and avoiding the subsequent kicks. Also cows have four legs not two which means the Blue Heelers are expert at avoiding kicks.

(4)   If you do lash out and attempt to kick a Blue Heeler you run the subsequent risk of pulled muscles, a strained groin or even worse the Blue Heeler could front up for more. Nothing in this kit can save you from the possibilities mentioned in (4).

(5)   Have a happy day.

Chapter 26

WATER WORRIES

By mid 1986 things were starting to move with contracts let for the supply and delivery of the 108 km of pipes to deliver the borefield water to the desalination plant. Tubemakers were awarded the contract in two parts. Their 375 mm diameter ductile iron pipe would be used for about half the length from Gab 6 to Canegrass. Sintercote pipe of similar size would be used for the run from Canegrass to Olympic Dam. This is a spirally welded steel pipe protected on the outside with a bonded polyethylene coating. Both types of pipe would be cement lined with spigot and socket joints sealed with rubber rings.

The Borefield Road soon became a major trafficway as the water tankers were joined by the heavy vehicles delivering pipes. The sides of the road were soon dotted with neatly stacked heaps of pipes at regular intervals.

Max Wallace, a small Canberra contractor, was the lowest tenderer for the work of constructing the pipeline, in competition with the majors. He was awarded the contract on condition that the Blue Heeler had to be left at home. Max soon had plant on site starting at the far end at Gab 6 and working towards Olympic Dam. He had gathered a team of drought stricken dairy farmers from the N.S.W. south coast to supplement his small permanent workforce. The team elected to stay in the Olympic Dam camp. In no time they were a co-ordinated team making good progress in spite of travelling up to 100 km each way each day.

They had excavated the trench,layed and backfilled about 18 km of pipeline and were approaching the Gregory Creek crossing when once again nature stepped in. It was another heavy rainfall closing the road to traffic for a few days. Worse than that Gregory Creek was again in full flood and the plant was trapped on the northern side. It would not be possible to continue the pipelaying across the creek bed until well after the flow stopped. The pipes at the crossing were to be two metres below the surface so the ground would need to dry out fully before digging the trench. As soon as the road was dry the contractor managed to ford his heavy plant to the south side of the the creek. Pipelaying was resumed leaving the gap to be filled in later. The water tankers worked overtime to make up for the time lost when the road was closed.

There was also a lot of activity at the borefield. Our groundwater consultants had contractors running seismic traverses to help in selecting the most favorable sites for production bores. A small diameter observation bore was first drilled and tested at the selected site. This provided the characteristics and thickness of the aquifer and the safe yield at the site. This was followed by the design, drilling, developing and completion of the production bore ready for later installation of a down hole borepump. Five new production bores were completed. These with Gab 6 would safely supply the 9 megalitre daily demand. The observation bore at each production bore would be retained for regular monitoring of drawdown.

At Gab 6 other contractors were busy preparing the site for the pumping station that was to push the water the 108 km to the mine. As well as the mainline pumps there would be a diesel engine driven power station to provide an electricity supply for the bore pumps. It was not economical to run a high voltage line up from Olympic Dam so we had to generate power at the borefield. Overhead lines were built for supply to each bore and the powering of the electric bore pumps. Underground pipelines were installed to deliver the flow from the bores to a large storage dam at Gab 6. The storage enclosed by earthen embankments was to have a plastic liner and floating plastic cover to prevent contamination. The storage inflow was to be controlled automatically by float switches operating the borepumps through control cables installed with the pipeline.

The traffic on the busy Borefield Road was now joined by trucks bringing in concrete for the foundations for the pumps, power plant and pipeline anchors.

In the last months of 1986 it all started to come together with Max Wallace and his team making excellent time with the pipelaying. On the last full day of laying the ductile iron pipe they installed a record continuous length of just over two kilometres. They drew closer and closer to Olympic Dam as the summer approached and were constantly working in temperatures of over 40 degrees. As soon as the section at Gregory Creek was completed progressive filling, scouring and pressure testing of the line started. The pipes were tested at well above the operating pressures and the only leak detected was at one faulty pipe collar near Gab 6. A small engine driven pump was used for the initial filling of the 108 km of pipeline and to start the flow to Olympic Dam in Febuary of 1987. That ended the constant tanker traffic that had sustained the project from Woomera and the Artesian Basin since 1980. The incoming water was sufficient to maintain supply to the small covered storage feeding the desalination plant. Any surplus was added to the reject brine from the plant and used for construction purposes.

The 400 kilolitre per day plant was keeping up with the fast increasing demand for potable water following the completion of the first houses in the town. A much larger plant using the Reverse Osmosis means of desalination was soon to be completed alongside. Two large clay lined earthen storages were under construction immediately to the west. The first, a 120 megalitre process water storage for supply to the metallurgical plant would receive all

Water Storages, Desalination Plant and start of Borefield Road at Olympic Dam.

Failure – At Process Water Storage.

the pipeline flow except that going for desalination and also the reject brine. Abutting this a 60 megalitre potable water storage would receive the product from the R.O. plant ready for distribution to all areas. Supply to the mine and office and metallurgical plant areas would be by direct pumping into mains at

After the Failure – "Where has all the water gone?".

The Water Storage – The Escape Route!

each location. A separate pipeline would feed to the village and into the 10 megalitre lined and covered storage at the town. The outlet from the town storage was the point where water would be metered for delivery to the Municipality and to a pump station pressurising the town mains.

There was a major hiccup when the installations at the Gab 6 main pumping plant and the bores were completed and as the 120 megalitre storage began to fill. It was just over half full one afternoon and less than an hour later all of the 70 million litres in it had disappeared without warning. Two forlorn looking visiting black swans were standing in a small puddle that remained. The desalination plant operator had climbed the perimeter embankment to check how the filling was going and was confronted with the empty dam. He raised the alarm so the experts could gather and consider the position.

There was a gaping sand sided hole about four metres deep in the floor against the sloping northern embankment. From the bottom a crawl sized tunnel sloping down had carried the water in under the bank. It was not safe to crawl far into the tunnel but it was obvious that it led to a natural shaft in the limestone deep under the surface. A swift calculation indicated that it must have drained the pond at over twenty cubic metres a second. The other astounding fact was that there must be caverns in the limestone at the bottom of the shaft capable of holding more than the seventy thousand cubic metres of water that entered in such a small time. It was a spectacular engineering failure which left all of us with very red faces.

As soon as plant could work on the floor a start was made to excavate the now loose material around and from the bottom of the hole. It was lightly cemented sand, with a lot of it richly coloured with ancient iron staining of varying hues. It was soon a very big hole as the loose material was removed but the shaft remained out of reach far under the bank. Eventually the pond floor and bank was reinstated with solid compacted fill and the decision made to seek funds to fully line the floor and walls with a waterproof plastic liner. The clay liner in the adjoining potable water storage also showed signs of leakage so this was also lined with potable quality plastic.

The theories of what caused the monumental failure came thick and fast. It was obvious that there were weaknesses in the clay lining but I believe the proximity to the inlet pipeline gives the clue to the rapidity. The ductile iron pipe in several lengths with rubber ring joints was set in the sloping embankment. It entered a cast iron bend when it reached the floor then continued horizontally to an inlet pad some distance clear of the bank. My opinion is that the increasing weight of water on the floor caused it to compress the sand underlying the clay lined section carrying the inlet pipe. The settling and small downwards movement in the floor would take the horizontal pipe with it. This would be enough to to open the rubber ring joint where the sloping section entered the bend. The jetting effects of the water pouring from this open joint would quickly scour a hole through the clay liner and the underlying sandy material would quickly saturate through to the mouth of the shaft and slump into it. The contents of the pond would follow almost immediately at a rate governed by the size of the shaft and the capacity of the cavern.

It was a lesson for me as I had selected the site for the storages and approved the drawings and details for tendering and construction. In this limestone country, there appeared to be no way of economically predicting the presence of a shaft into the limestone so far under the surface. Seismic testing might have indicated the cave system which may or may not extend under the storages at depth. The pond floor was in excavation at about four metres below the surface and generally with a solid gravelly limestone base. In retrospect the underlying sandy material where the inlet pipe entered should have indicated we were near an ancient sand filled sinkhole or similar. In any case we certainly improved the quality of the saline groundwater in this limited area.

The two storages, complete with their plastic liners were eventually filled in time to safeguard uninterrupted supplies to the mine and town. Meanwhile the other works to get the project up and running were proceeding with increasing numbers of contractors arriving on site. There was some urgency to complete the 132 KV power line which would run from the State power grid at Woomera to a project switchyard at the plant area. Doug, Peter Hill and I arranged a meeting with the Administrator at Woomera to finalize details as to the point of connection and other matters.

We set out from Olympic Dam in a hired Land Cruiser in plenty of time as the road was drying out from recent rains and greasy on the gibber section. There were no problems until we reached Phillip Ponds where there was a section of the road covered with six inches of muddy slush that had been churned up by the heavy traffic. It was quite solid underneath and we were making steady progress until a blowout in a rear tyre stopped us in our tracks.

It was right alongside the ruins of the old Phillip Ponds Homestead that seemed to still be cropping up in my life. The three of us were nicely dressed with coats and ties for our meeting at Woomera. Short of walking one of us would have to get out in the slush and change the tyre and even then we would be late into Woomera. After 39 years since my first time at the Ponds I was still outranked. I got out, and kneeling and slopping around in the mud changed the tyre with great difficulty. Peter and Doug had made it to higher ground beside the road and freely offered advice and encouragement.

We were very late, and muddy to varying degrees when we finally made it to the understanding Administrator's office, where fortunately the location for the connection to their power line and for a switching yard to the south of Woomera was agreed. The prestressed concrete poles for the power line were made at Port Augusta and the construction contractor soon had them streaking across the gibber in perfect line on the run to the mine. Prior to completion the power plant at the mine managed to meet the increasing loads of construction and supply down the new 33 KV pole line from the village to the town. There was relief when the State supply was connected at a project switchyard to the north of the plant area in September 1987 and the diesels were put on standby.

Chapter 27

BACK AT THE MINE

The office adjacent to the airstrip was stretched to capacity meeting the increasing operational requirements pending completion of the new main administration office and store. The building was completed in mid 1987 and Bob Crew and his staff from the airstrip office moved in. They were quickly followed by a number of others transferred to site from the Adelaide office. The solidly built transportables from the airstrip were then relocated to a site behind the new building office. These became the Engineering Office for the site maintenance section now headed by Bill Rymill.

There were major new works to be carried out at the Whenan Shaft in preparation for bringing it into full production. The decline road access from the surface to the underground workings was completed in mid 1987. The heavy mobile plant could then be driven in and out of the mine instead of being lowered down the shaft in pieces. They could now be readily brought to the surface for repairs and servicing. A large workshop building was planned to supplement the underground facility. There would also be a new mine office, a laundry for handling work clothes,and new change rooms for the greatly increased workforce and the many visitors warranting an underground visit.

The core farm was to be moved from the village to an area south east of and adjacent to the shaft site where a new sample preparation building would be built. Once again adequate provision for drainage and collection of stormwater received attention as it was in a small closed catchment.

The shaft would be equipped for hoisting all the ore from the producing mine and would be a centre of activity. We had set about the detailed planning for the installation of permanent services and road construction and sealing as a matter of priority. The work had to be carried out with minimum interruptions to the operating mine and required meticulous attention to detail and timing of the various works. Eventually the obstacles were overcome and Kinhill had contracts underway for the roads and services. At the same time Mike had the ore handling works at the shaft underway and Fluor were preparing plans for an overland conveyor to take the ore from the mine to the plant.

The existing road access for the workers at the mine would be cut off by the security fencing that surrounded the whole plant and mine area. A new road had to be built bringing in the traffic in a semicircle around a large waste rock stockpile to the east of the mine. The road ended alongside a large car park and adjacent to the controlled entry into the shaft site. A short connecting road would provide access to a viewing area on a dune overlooking the decline entrance and shaft site. This would cater for the increasing number of visitors interested in the project.

The Environmental Officer, Barry Middleton, and his growing number of specialist offsiders and consultants were also taking a keen interest in our construction activities. It was decided to build them an office with laboratory facilities and fenced yards to house them and their equipment, kangaroos, birds,snakes and lizards. The site selected was in the village on the town to mine road, now called the "Olympic Way." That would have the benefit of being accessible to visitors who were always interested in our attention to environmental matters.

Barry and his crew, were active with detailed studies of the flora and fauna, and at the artesian basin mound springs. They also kept looking over the shoulder of those of us on construction. It took Hughie some time to convince them that the new section of Olympic Way, the main road from the town to mine, could not be a series of tight bends to miss every tree.

By the end of 1986 the weeks routine of for Doug, Mike, Hughie and I was becoming repetitive. Mondays and Tuesdays were long days in the office with meetings, checking drawings and specifications and costs against budget allowances, more meetings, and setting priorities. At 6.25am Wednesdays saw us racing across the airport tarmac to claim our regular seats at the back of the aircraft for the trip to Olympic Dam. This was very important as the Metroliner aeroplanes on the run were extremely noisy. Ear plugs became a must unless you were seated right near the tail. On landing Bluey had one of his Budget hire vehicles waiting at the airstrip for those of us who had the need. I usually picked up a 4WD as the works in my domain extended from Woomera to the borefield. First we would all gather at the construction office for our weekly meeting with the on-site construction supervisors. The afternoon saw us checking on the works in progress, Doug, Mike and I in the mine and plant areas and Hughie at the camps and Town.

We would usually make it to the wet canteen by 6.30 to socialise with the locals and to the mess for dinner as they were closing the door. After the meal and settling in to our camp units we would usually find ourselves in the smoke filled lounge of Hughie and Winston's unit. This was the relaxing part of the day usually helped along with plenty of visitors carrying a few stubbies, and lots of talk. Doug had only arrived to head the project development a few months before construction started. We were a bit unsure how to take him and obviously we were unknown quantities as far as he was concerned. In a very short time any possible hang-ups and differences had disappeared. It was the ideal situation to let off steam, get to know each other socially and build up a genuine mutual respect. Hughie and I were the old men in a young persons construction environment but got by when we told the others we still remembered what it was like to be young.

The mess only served breakfast until 7am so it was another early start on the Thursday morning. This had been the opportunity to travel the borefield road, to inspect the progress on developing the new bores, the headworks at Gab 6 and the pipeline contract. Back at Olympic Dam there was the water and power reticulation and site works at the mine and plant. At 3.30 it was back on the aircraft for the return trip. Fridays it was a quick sort of the accumulated paper then Doug's regular project meeting with the Kinhill and Fluor design, contracts, and costing staff.

I managed to get some short breaks from this routine with a number of trips to Perth, Kalgoorlie and Norseman. George White now in charge of exploration at the historical Bendigo goldfield also had me visit with Cec Forbes to advise on dewatering of a line of old workings.

Chapter 28

THE TOWNSHIP

The town and village were now officially in the Municipality of Roxby Downs which immediately came into being when the Joint Venturers gave notice to the State that the project would proceed. John Harris was nominated by the State as the 'Administrator' to manage the Council affairs during the initial construction phase. A permanent appointment would be made as the town grew and the facilities became available. Later on an elected Council would assume command. Under the terms of the Indenture, the Council in addition to it's normal role, was to be the water, sewer and electricity authority in its area. The Company had the obligation to provide power and water to the Municipality in bulk at subsidized prices.

Town Area –
Before Development

Clearing for the First Town
Streets – 1986

The Town –
The first houses appear
– 1986

The Town Caravan Park
and Housing – 1987

The Town Community
Club and School – 1987

One of the first contracts let was the fencing of the agreed municipal and mining lease boundaries, a total length of over 92 km. A further section was added within this area to enclose the townsite and surrounds with a wire netting fence so that area could be cleared of rabbits and foxes. Hopefully that and the exclusion of stock from the remainder of the fenced area would result in the rehabilitation of the native vegetation and reduce windblown dust. Geoff Witham met with the pastoralists and made sure that they were adequately compensated for the loss of their grazing rights.

The town construction got under way with access from the Woomera Road firstly along the station track and then by forming Axehead Road immediately to the south. The town streets were then progressively cleared and the site for a town construction office and yard prepared. The office was soon assembled from a number of Atco units. Hughie had insisted that Brian McKay be the onsite Manager for the works at the town an he was now full time at Olympic Dam. Assisting him were a number of Kinhill construction inspectors including Brett Allen returning after seven years.

The Village Industrial Area was expanding with sites being developed by numerous organisations servicing the mine and plant construction. This was to be the 'heavy' industrial area with serviced allotments being provided for the concrete plant, earthmovers, plant hire, repair shops and other users. Only Light Industrial sites would be provided in the township, in an area on the northern side of the town adjoining the construction office. Sites were quickly surveyed and made available to the town building contractors, service industries, the electrical switchyard and the gas storage tanks. The road along the western boundary of the town, Olympic Way, would be extended on to the north as the town to mine road. BP selected a site fronting this road to develop as a service station for the town.

The Company Environment Department had equipped themselves with a brand new heavy duty, you beaut, 'wood chipper' to produce material for landscaping use from the trees and shrubs being cleared from road and house sites. They eagerly lined it up alongside the heaps of material that the road builders had placed in stacks for them and started it up. The dense hard wood of the mulgas and myalls was too much for it, ruining a set of cutters in a day, and the machine in two.

The Readymix Quarry and crushing plant were now in full operation about 3 km west of Olympic Way along the Axehead track. A road to bring out the urgently needed crushed road base and aggregates for the mine and town, and also construction water from the bore became a priority. It hadn't been included in any of the existing contracts. Hughie and I soon found out it was no longer a matter of pegging the line, writing an order, and having a grader and trucks build it in a week at minimum cost. We now were bound by the requirements of a 'Procedures Manual' with rules and conditions that defied description. To get the road trafficable according to the book would cause long delays and at least double the cost. In the end we bucked the system, had the road open in quick time at minimum cost, and the rules were later modified to allow us some minor discretions without blowing up the computers.

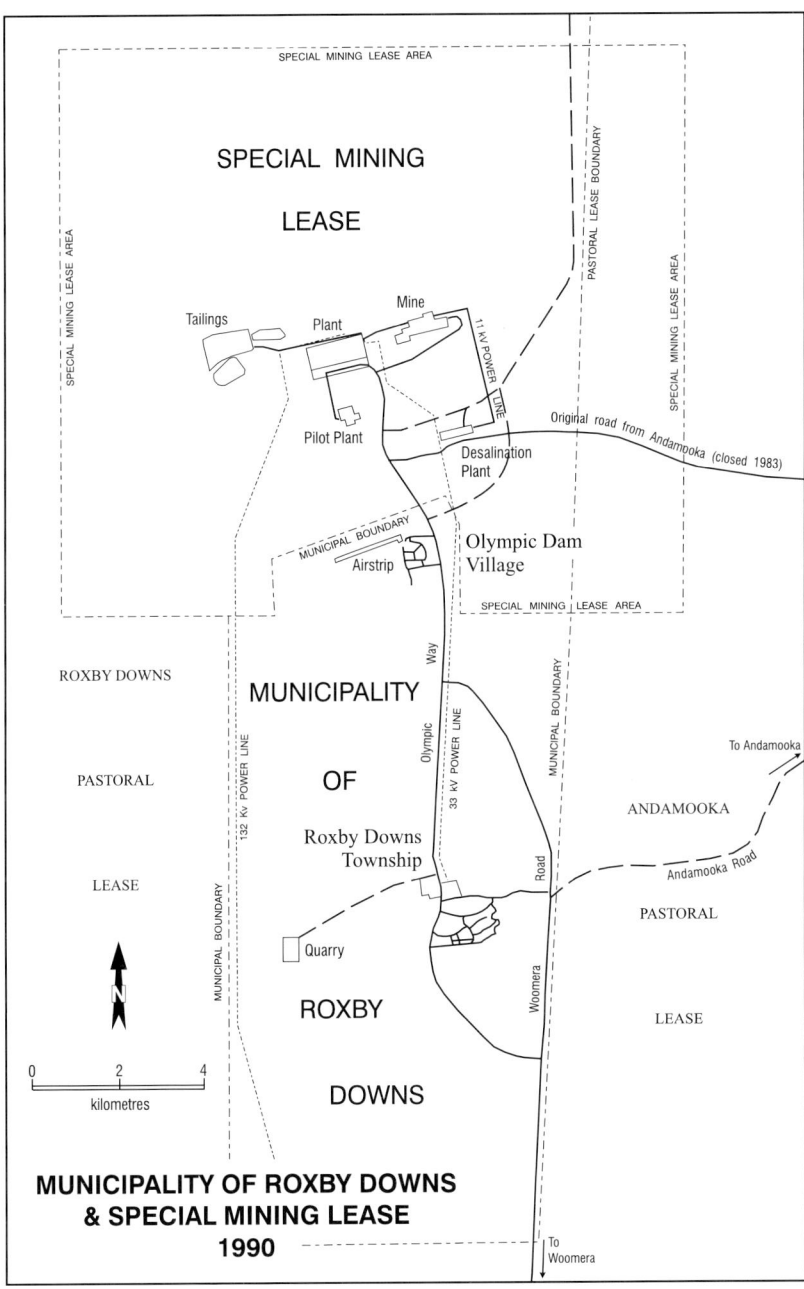

SPECIAL MINING LEASE AREA

SPECIAL MINING

LEASE

PASTORAL LEASE BOUNDARY

SPECIAL MINING LEASE AREA

SPECIAL MINING LEASE AREA

Tailings   Plant   Mine

11 KV POWER LINE

Pilot Plant

Desalination Plant

Original road from Andamooka (closed 1983)

MUNICIPAL BOUNDARY

Airstrip

Olympic Dam Village

SPECIAL MINING LEASE AREA

ROXBY DOWNS

MUNICIPALITY

Olympic   Way

132 KV POWER LINE

33 KV POWER LINE

MUNICIPAL BOUNDARY

PASTORAL

OF

Roxby Downs Township

MUNICIPAL BOUNDARY

To Andamooka

ANDAMOOKA

LEASE

Andamooka Road

Quarry

PASTORAL

N

ROXBY

Woomera   Road

LEASE

0   2   4
kilometres

DOWNS

**MUNICIPALITY OF ROXBY DOWNS
& SPECIAL MINING LEASE
1990**

To Woomera

By the end of 1986 Hughie had housing contracts for 120 houses and the single quarters complex well under way. The construction of a Licensed Club building to Max Hannel's design and financed by the company was also in

Hughie and I – Off to Work

Wednesday Night during Project Construction. Hughie, Author, Doug and Graham.

progress and completion was eagerly awaited. Installation of roads and services were nearly keeping up with all power, gas, phone and water lines underground in the street verges. Only street lighting poles and signs were above ground, and tree planting would follow later.

The Town Centre

I gave priority to the construction of the power and water lines to supply the town from the mine area. Telecom was giving excellent service and had their communications tower on a high dune to the north of town up and in service quickly. This added to the incoming lines as an interim measure and with cabling installed from the town to the mine gave some service to all areas. The new camp in the village was soon filled to capacity catering for the continuing build up of the construction workforce.

Hughie was having problems on the house construction with the inspectors having difficulty in maintaining standards. He was desperate to find someone with a depth of experience and authority to deal with the building contractors and to give backup to the inspectors. He finally came up with just the answer. We had played golf most Saturdays with a nearly retired builder who had managed his own successful building company for many years. Graham Shelton was also a 1926 model who's only fault was that he played golf left handed. Anyhow when approached 'Leftie' said he would give it a go for a couple of days a week and soon had the building works in the town under control. He joined in our two day midweek trips to site including the gatherings on Wednesday nights.

Graham helped out on a part time basis on the town works towards the end of the construction period enjoying every minute. During that time Bob Crew had noted his ability and character and offered him a permanent position getting the site building maintenance organized and managing it. He got approval from his wife, Yvonne and they moved up to Roxby Downs to a newly completed house as one of the early residents.

The State Public Buildings Department had prepared the plans and specifications for the Education Complex, and Hughie had tenders called and contracts let as a matter urgency. The construction of the adjoining Library, Auditorium, Indoor Sports Centre and Swimming Pool would follow. These were the first of the facilities to be provided by the State with the construction contracts managed by the company. They were all fronting the northern side of the wide main street in the town centre. Behind them and to the west on the same block of land were to be grassed Playing Fields for the School, Tennis and Netball Courts and the Sports Oval. Opposite and at the western end of the street would be the Police Station, Medical Centre and Ambulance Station, and combined Municipal and Government offices. Further to the west a site was set aside for a future Hospital.

At the south eastern end of the the town centre sites were set aside for a Hotel-Motel, Tavern and Bottle Shop, and a Shopping Centre. Hughie, Doug and John Harris had spent a great deal of time seeking private developers prepared to finance, build and operate these commercial facilities in the town. This was made difficult as the mine production was to start up at a considerably lower rate than earlier indications. There would therefore be a smaller population in the town and potential developers had to have faith in the certainty of later expansion.

The village now had a mini-supermart, a post office and a post code - 5725. These with trips to Woomera and Port Augusta would serve the shopping needs of the town and village until there were shops in the town. Bluey and Jan's thriving service station and repair shop covered the motoring and hire vehicle needs in that area.

Lengthy negotiations eventually led to a start on the building of a shopping complex fronting the divided roadway in the town centre. Alongside the development of an imaginatively designed Hotel-Motel and a separate Tavern-Bottle shop-Diner also got under way. The Hotel and two storey Motel buildings enclosed a courtyard which was to have a high central mast and a conical white sail roof. When completed with swimming pool and tropical planting and landscaping it became an shaded oasis. The high white sail became the prominent landmark in the developing townsite, and visible in the distance from approaching aircraft.

The grassed areas in the town including the Oval, Playing Fields and a central plantation in the main street would be maintained using reclaimed water. This would be mainly treated sewerage effluent supplemented with stormwater harvested from the infrequent rains.

The first houses were completed, services connected, and occupied in Febuary 1987, and the Licensed Club was opened in style at the end of March. Those heading the priority list for the first houses were mostly long time residents from the village van park. They lost no time in moving in. This and the occupation of the new caravan park in the town partly relieved the pent up demand for serviced van sites by the increasing numbers of contractors. A quick build up of town population followed as more and more houses became available. Occupation of the Hannel designed Mess and town single quarters soon followed. These were on a prime site with some magnificent native pines, adjacent to Olympic Way and within easy walking distance to the club and town centre.

The Andamooka school was now catering for increasing numbers of children travelling on the school buses from both Olympic Dam and the new township of Roxby Downs. The original road to the Opal Field from the mine area had been closed off at the time of the demonstrations. It was replaced with an easterly extension of the Axehead track branching off the Woomera Road at the townsite. It rejoined the original road at about the half way mark near the 12 mile dam. It was still a dusty gravel road and the school buses didn't run when it was wet. It was a great relief for all concerned when the Roxby Downs school and housing for the staff were finally completed.

Others to become early occupants of new houses in the town were those long time residents who had been living in the Village housing area. They were probably a little sorry to leave the small close community that progressively grew to 16 houses after the first were occupied by the Whites, Crews and the shaft sinking supervisors in 1981. Bob and Lyn Crew were still there and had been joined by a number of those who had been on site since the early days of the project. Jack Clarke, a drill foreman had been an original occupant of the Olympic Dam camp when it opened in January 1980. Shortly after the Masons and the Perkins were the first to settle into the Village caravan park. Other long time residents in village housing with their families were Ron Mills, Ray Spiller, Tassie Compton, Bob Mundey, Rob Smith, Barry Hewlett, Frank Boulton, Bruce Sprott and Gordon Bowe. Later arrivals in residence were Jim Reeve who had replaced George as Chief Geologist, Mark Sonter a radiation specialist, Morrie Daw in charge of administration, and Bill Rymill who replaced Gerry Gould as the Chief Engineer.

As soon as the houses were vacated work started to prepare them for transport to new sites in the town where they were quickly re-established for new occupants. The power poles, fences and all signs of habitation were soon removed from the vacated settlement. Roads and house sites were then ripped and prepared for the re-establishment of the native vegetation and the area closed off to further disturbance.

In addition to the retention of all possible existing vegetation the planting of trees and shrubs throughout the township was always regarded as a critical part of the development. In particular the creation of the maximum amount of shade from trees suited to resist the extreme summer temperatures with a minimum water requirement.

As the town grew and became more settled Bob Crew thought it time to approach the Joint Venturers for money to establish a Golf Course and a Bowling Green. He asked me to come up with siting proposals and estimates of the funds required. It had always been in mind that the large low lying area with spaced dunes abutting the town to the west would be the site for the course. We had good contour plans of the area and I provided Bob with the detail and a hypothetical layout plan late in 1987. I had to brush up on my municipal experience to put together a detailed design and estimate for the bowling green. This proposal included a synthetic grass playing surface as water was too precious to maintain real grass to the standard required. Once again accurate levels and thorough drainage of the level surface would be essential. A site opposite the single quarters, adjacent to the oval, and across the carpark from the Club and Tennis Courts seemed appropriate.

Crewie must have persuaded the Management Committee of the necessity for these facilities as the funds were made available.

On site Bill Rymill and others quickly set about having the Golf Course layout refined and getting it prepared for play. I had allowed in the estimates for the collection and storage of the intermittent stormwater flows from the town outfalls. A large dam was excavated and pipes installed to deliver this water to the dam. Hopefully there would also be some surplus reclaimed water from the sewer lagoons after the needs of the oval and playing fields were satisfied. This at best would not water much of the course but would give the players some green patches to look forward to after playing on the bare dirt and oiled greens. It eventually became an interesting, if different course with the players each having a small mat to place their ball on for each shot. The dunes bordering a number of the fairways also formed exceedingly long sand traps to be avoided if possible, and there were also plenty of mulgas to trap the hackers.

When the time came to build the Bowling Green it was carried out under the direction of Graham Shelton who had local contractors carry out the work until it was ready to roll to his satisfaction. At the time the initial project construction was well advanced and there were a number of open floor space 'Atco' office units surplus to requirements. Graham gathered them up and refashioned them into very acceptable initial Club Rooms at both the locations.

By this time the town was really taking shape, the school and other buildings being funded by the State were progressively completed, staffed and occupied. The Swimming Pool with all facilities was completed for the

1987/88 summer and the surrounds decorated with tall palm trees landed at great cost and plenty of sail cloth shading. The Shopping Centre, Tavern and Hotel-Motel were built and opened expeditiously to serve the growing population. The sealed roads and streets and services were progressively handed over to the Municipality. John Brazel had been succeeded John Harris as Administrator of the Roxby Downs Municipality and resided in the town attending to council affairs.

The planting of trees throughout the township was placed in the hands of the State 'Woods and Forest Department' and much to Hughie and my distaste submerged the place with eucalypts. I had planted River Gums at the temporary office site at the airstrip and they certainly grew at a very fast rate with a minimum of care. These and other varieties certainly had a place in the plantings but many of the gums dropped a lot of litter and can become dangerous as street trees. The nearest eucalypts to the town were those lining the creek beds 100 km to the north and were not naturally found growing in the dune areas. We had set out originally to have a wide shaded plantation separating the traffic in the main street, This was copying the attractive towns in the South Australian Riverland, with their dense shaded areas from deciduous plantings for the fierce summers and filtered sunlight in winter. It was not to be.

Chapter 29

CONSTRUCTION

As the town grew so did the installations in the various metallurgical plants and in the services area. By the end of 1987 the administration office and stores buildings were occupied. Bob Crew as the Resident Manager had a prime office with a southerly outlook over a landscaped area and on to native vegetation fringing a claypan. We had asked a contractor to collect some attractive rocks for the landscaping which he promptly delivered. These were carefully arranged in a landscaped garden fronting the office. Unfortunately the location where the rocks came from was then pronounced to be a sensitive site and the rocks were reloaded and returned from whence they came. We considered what we could do to make up for our misdemeanor. In the dead of night a very nice garden gnome, obtained after many difficulties, was placed in a suitable spot directly under Crewie's window as a surprise for him. It was not appreciated and before 7 am it had disappeared never to be seen or heard of again.

A contentious item during the planning stages of the project was the best means of storing the tailings from the plant. The proposals in the E.I.S. provided for deposition of the tailings slurry progressively around the perimeter of compacted embankments. The inwards sloping beaches of deposited solids would then be kept damp and the surplus tailings liquor collecting at the centre of the storage decanted off to a separate evaporation pond. This was followed by further detailed studies and findings after the testwork at the pilot plant. A significant result from all this was that crusting of the surface would prevent dusting and hence no longer a need to keep the surface damp.

This subject was pursued at numerous meetings with the metallurgists, our specialist consultants and representatives from the Mines and other State Departments. Eventually the decision was taken, and approved, to dispense with the decant and to provide sufficient storage area for total evaporation of the surplus water. The sizing took into account the very low annual rainfall of 160 mm and an effective evaporation rate considerably less than the actual pan evaporation of about 3 metres.

The storage was designed and constructed on this basis with compacted and reinforced earthen embankments. Although about 800 metres square it took up only a small part of a large area set aside for this purpose a few kilometres to the west of the plant. Test holes had already been drilled to ensure the site was clear of the deep mineralised zone. These drill holes were also to be used as part of the overall groundwater monitoring system around the mine. The recording and plotting of levels of the saline aquifer was already showing increasing drawdowns sloping consistently towards the shaft.

Completion of the tailings storage was followed by the laying of the tailings delivery pipeline from the plant. This surface pipeline was contained in a narrow corridor with earthen bunds each side to contain any accidental spillage. The system was then ready to accept the tailings flow as the plant was progressively commissioned.

The first half of 1988 saw the results of the years of effort come together. The large surface stockpile of accumulated ore and that from the increasing underground development and crusher started to flow to the new metallurgical plant. The long overland conveyor from the Mine dumped the ore on to a high stockpile where it could be reclaimed on to another below ground conveyor and fed to the plant as required. There the commissioning phase and final checks on all the new installations and equipment was underway. The roads throughout the plant areas were edged with kerb and channel and sealed to ensure ease of cleaning and a dust free site. The large steel framed buildings for the Concentrator, Uranium Plant, Copper Smelter and Refinery were at the final stages of completion. The individual specialised items of plant were being progressively installed, tested and commissioned. At both the Mine and the Plant the increasing numbers of technical staff, mainly from other company operations, were settling into the newly finished offices, control rooms, and underground areas. Other than the nucleus of experienced operators most of those recruited to work in the metallurgical area were unskilled. A large number of these were from the depressed rural areas in the States' Peninsulas where years of drought had undermined the farming economies. The opportunity of assured well paid employment with new houses or single accommodation provided at low cost was very attractive to the young sons of the land. As it happened it was also for the wives as there was now the increasing tendency to employ females in all areas of the workforce. The isolation of Roxby was not of major concern to these rural people as they were well used living far from the joys of the city. All staff and employees had to undergo an extensive familiarisation programme outlining the company policies ranging from environmental matters, work practices to radiation protection and monitoring.

In the town the final touches were being given to the last of the civic buildings and recreational facilities to be completed. The permanent Single Quarters were in full operation as was the Hotel-Motel, Tavern and Shopping Centre. The school was a hive of activity and gardens were flourishing on the sandy surrounds of the growing number of houses. There was now ample water in all areas with the completion and commissioning of the new 4 megalitre a day Desalination Plant. It was even possible to accede to a request from Andamooka to draw on the supply to a limited extent.

The build up of reclaimed water had enabled the oval and playing fields to be grassed. Local sporting teams were competing with each other and with some from Woomera. The Telecom exchange linked to the outside world by optical fibre was officially opened with due ceremony in Febuary. The ABC television signal and FM radio was available from Company funded transmitters on the high mast on the prominent dune to the north of the town. A friendly technician had managed to add our original village transmitter to relay the AFL football from the satellite to satisfy the urgent demand in the previous September. Hughie was busy signing the hand-over certificates for the continuing house building program. In the occupied houses there was a mushrooming of private above ground pools. The Municipal Pool was also very popular until the cooler late autumn weather. Tree planting in the newly completed streets was showing results with the native vegetation adherents still in full control.

By now there were scheduled air services into Olympic Dam as well as regular charter flights coming and going. The Air Force even tried it out by landing a couple of Hercules aircraft. The original natural gravel strip was extended in 1980 and a short section at the east end sealed. It was widened for licensing in 1982 and had handled the increasing traffic with minimum maintenance. As part of the project development the rest of the runway and an extended apron area at the terminal were sealed. The terminal facilities were minimal with a small open roofed area joining a small toilet building to an equally small check-in office.

The whole of the project construction was carried out with all materials and equipment transported in on the original inexpensive dirt road from Woomera that we had built years before. As well as the construction traffic there was now increasing numbers of furniture vans, the supply vehicles to sustain the large population, and the coming and going of the residents and visitors. The surface was well maintained by the Olympic Dam engineering maintenance group and was regularly watered and graded. In spite of past predictions the surface had never developed corrugations. Other than for short periods after rains it had been a high speed road giving no trouble and had been largely accident free.

The Highways Department had been ready to go on constructing and sealing the new road from Woomera at our cost for some time. Hughie and I being used to building low cost roads in rural Victoria thought that their estimated costs were far too high. Doug sided with us, regular meetings with the Department led nowhere, and the Joint Venturers held back on formally requesting the construction. In early 1988 we accepted the inevitable and funds were made available to give the go ahead.

The contract for construction was let to MacMahon Contractors who were no strangers to the area. They had previously completed the major site earthworks, roads and services installation contract for us in the metallurgical plant area. Other contracts were let for the supply of the massive amounts of crushed rock pavement material that was needed.

The material for the northern end of the road would come from the Readymix quarry at Axehead and was immediately available for a start. A new quarry would be opened up by another operator to the south of Purple Downs Homestead and would be produce the pavement material needed from there to Woomera.

MacMahons were quick to get going and started at the Mine Lease boundary about a kilometre or so north of the village. We had previously resurfaced and sealed our road from there to the mine, plant and office areas. The new construction progressed south on top of our existing road into the village. They then rebuilt the short spur into the airstrip and camp before continuing on. It was new construction past the east side of the village and half way to the town to meet up with Hughie's Olympic Way. From there the new road would closely follow the existing one all the way to Woomera, bypassing the town at a distance to the east, The Indenture requirement to provide access into the town was met by extending Olympic Way to the south from the township to link in with the new construction 4 km further on. As the work progressed, the residents of the town and village could readily assess the progress as the new roadworks were generally less than a stones throw from the original.

In July as construction drew to a close Doug arranged a "Break Up" party before the contract workforce and supervisors finished the last of the work and left the site. It was held in the large dining hall of the main construction camp which filled to capacity. Hughie who had finished his contract and returned to his home in Bendigo came back to Adelaide ready to fly to Olympic Dam to attend. Unfortunately he became ill and didn't make it back to join in the celebrations.

Most of the construction I had been looking after at Olympic Dam was now complete. When Hughie left I was sharing my time on matters affecting the town, working for George White at Bendigo, and investigations relating to the phosphate deposit in north Queensland. One task in connection with the Roxby Downs township was the assessment of the value of each and every allotment and spare piece of developed land. Our costing system made it easy to determine the total expenditure on land development in the town boundaries.

The Indenture required that only the amounts actually incurred in providing internal roads and servicing could be used in arriving at a value for each individual piece of land. The Company was obliged to provide power and water to the town boundary at their cost as well as that of incoming Telecom services. The allotments for Government facilities and housing, for private development or for sale all had to be made available in accordance with an agreed apportionment. Only a small area of the total townsite had been occupied and it was obvious that the undeveloped areas should bear some of the costs. This was particularly so where the roads and services were in position for easy servicing of adjoining land when the time came.

The initial assumptions and apportionments were subject to progressive review and modifications entailing detailed recalculation made for each change. Fortunately Kinhill came up with a computer programme that took all of the hackwork out. Eventually a list of all allotments and the details of how we had determined a value for each were submitted to and thankfully approved by the State authorities.

This then enabled us to arrive at total "house and land" cost for each of the company residences and to review the options for an employees home purchase scheme. Individual home ownership was considered to be a desirable policy to 'normalize' the town both for the company and the employees. It would lead to a gradual lessening of company involvement in town affairs even though it would be the dominant employer. There were numerous meetings on the subject and just as many different opinions. There was a substantial ongoing cost to the company for maintenance, rates and insurances on all the houses that were rented to their employees. The low rentals that had become the norm at the inland 'mining towns' did not even cover these costs. I argued that the company could therefore make their houses available for sale to employees at a substantial discount and well below the actual cost. The assured long term life of the mine and a low price should ensure that purchase was an attractive option for the occupants. It was recognized that any home purchase scheme had to be on favorable terms to the purchaser with safeguards for both parties. A suitable scheme was eventually adopted with the hope that Roxby Downs would become a totally 'open' town subject to the normal forces of supply and demand.

By September the town population was becoming settled after the disruption of construction and settling in. Then came the opportunity for many of the new occupants to attend a unique function. The "Opera in the Outback" was being staged near Beltana (Population 7) on the other side of Lake Torrens as one of the more unusual Bi-centenary events. By outback standards it was a relatively short drive up the Borefield Road, then east to Maree and south past Leigh Creek to Beltana, a total of 300 km.

It was an occasion to camp out in the areas set aside with bare minimum facilities by the organizers. Those attending came from all over the country and beyond. There were 7 km of trains made up with railway sleeper carriages and dining cars parked on the nearby rail line which usually only carried coal from the Leigh Creek deposit. These catered for patrons attending from afar, with all the comforts of home on their excursion to and from the outback. The Roxby Downs visitors camped out with the hundreds of others ready and willing to 'rough it' in the areas set aside in the bush by the organizers. The camping facilities as provided included hessian screened toilets which caused much hilarity, when, from certain directions, all occupants were silhouetted by the powerful area floodlighting.

I took a few days off to drive the 600 km from Adelaide the day before the Opera, and early enough to pick a beautiful campsite in a shaded gravelly creek bed. (Thank goodness it didn't rain.) Somehow the Sheltons from Roxby Downs and other friends managed to find us among the spread out camping sites and settled in alongside. Saturday afternoon was a day at the Beltana races at the freshly graded dusty race track prepared for the occasion. The sun shone and all were intent on enjoying their time in the bush. As night fell buses shuttled the audience from their trains or camps to the entrance of the concert venue. It was then a half kilometre walk for the 10,000 attending. The track led along a dry creek bed edged with vivid floodlit white gums. Brick red, bare stoney hills rose steeply on each side to craggy ridges,closing in then opening into a spacious amphitheatre. There, the open air seating for 10,000 was carefully arranged facing a large roofed stage for the symphony orchestra and the star attraction - soprano Dame Kiri te Kanawa.

Following the the performance, there was a bush catered supper in the floodlit open spaces with small groups of folk musicians entertaining. Then the longish walk back to the camp with a nightcap around a blazing campfire.

Chapter 30

THE GRAND OPENING

The official opening of the initial Olympic Dam operation was programmed for Saturday November 5th 1988 when the mine and plant were in full production. It was to be a big occasion with invited guests coming from near and far. As well as those who had been closely associated with the project and the Joint Venturers, there were a large number of interstate and overseas visitors. These were mainly from organizations who were past customers or potential users of the metal products of the Joint Venturers. Then there were the politicians with the official opening in the hands of the State Premier, the Hon. John Bannon M.P. The Labour Party who had voted against the Indenture which enabled the project to proceed were now in power. Their Upper House member who had had defected from party policy and enabled the Indenture to become law was also to be an honoured guest. Norman Foster was no longer in Parliament, having been expelled from the party for not voting with them and defeating the bill.

Heading the numerous Company representatives were the W.M.C. Chairman, Sir Arvi Parbo and for B.P. their Managing Director Mr Patrick Gillam. The guest list had to be limited because of logistics of transporting large numbers to the site and the facilities available once there. I was a bit disappointed that a place couldn't be found for Rod Everett. It was his unrecognised intervention in 1979 that enabled the siting of the Village and later the Town in the, then, Woomera Restricted Area. The alternatives don't bear thinking about.

Bob Crew and his staff in their usual thorough way ensured all arrangements on site were checked and double checked. 'Olympic Dam Marketing Pty Ltd,' the Adelaide based sales group now formed to sell the project products handled the city arrangements. The overseas and interstate visitors were programmed to be in Adelaide on the Friday where they were booked in at city hotels. There was an official dinner at the Hyatt hotel for them on the Friday night. I had a phone call from Bob Crew some weeks before the function to see if I could contact Len Beadell and see if he was available to be a guest speaker at the dinner. Len having had a number of books of his outback exploration and adventures published was now a famous figure. He was also in demand throughout the country as a speaker at various functions. I made contact and he agreed to ring Bob Crew and make the final arrangements.

The big night arrived and we made a point of arriving early to meet Len and his wife Anne on their arrival to guide them among the crowded throng.

They were no doubt pleased to see someone they recognized and we quickly settled at an empty table. Len and I then went to check that the slide projector and microphone etc were all in readiness for his talk. When we returned to our seats at the table we found that the other four seats had been taken by Sir Arvi and Lady Parbo and two high ranking police officials. I immediately thought we may have picked the wrong place but Arvi quickly assured me that all was well and they were happy to join us. I had known him since the early days when he often visited Kambalda. He had also on occasion appeared at my Adelaide office door unannounced when he visited unexpectedly and was looking for someone he knew.

The dinner went well, Len gave his talk covering his outback adventures and showed his impressive slides of the colourful remote outback scenery and inhabitants. It wasn't a late night as there was to be an early flight on the way to Olympic Dam in the morning. As it happened it was the last time I was to see Len.

The large number to be flown in from Adelaide for the opening meant that there were more aircraft involved than could be parked at the Olympic Dam airstrip. The runway was now sealed but to insufficient width and length for the larger commercial jets. The few flying direct had to be content to arrive in the smaller turbo-prop aircraft. The overflow in the large jet aircraft were landed at Woomera and bussed to site. Everyone was issued with a folder with a coloured name tag, and a detailed itinerary of the arrangements made for them for the day. There was a holiday mood on landing and making our way to bus marked with the colour on our name tag.

It was a glorious inland, arid area, shirtsleeve, day as forecast without a cloud in the sky and no wind. The fleet of buses made their way to Olympic Dam on the original dirt and gravel road still in good condition and sustaining the project.

(It was probably the first time some of the overseas visitors had seen the outback and a dirt road, let alone travelled on one.) The trip was smooth and fast, landing us at the Environmental Centre in the Village for morning tea. The staff had ensured that there were some excellent displays of interest particularly for those in the desert for the first time. These included many of the local snakes and lizards safely behind glass and kangaroos recuperating from accidents on the roads.

It was then back to the buses for a tour of the shaft and plant area, then to the huge marquee where the opening would take place. We took a different course and went on an underground inspection with a small group allowed this privilege. It was a quick change in the visitors change rooms, driven down the decline entrance to the enormity of the system of underground tunnels and cavernous areas. Before noon we were back on the surface and joined the throng at the marquee.

This had been erected on a vacant area adjoining the operating smelter and as well as the official dais had neatly arranged seating for about 650 guests. They were soon all in attendance and seated in their colour coded seats and the ceremony began.

A small group of musicians alongside the stage led all attending in the National Anthem helped along by the words of the two verses printed on our programs.

Sir Arvi then gave an address followed by Mr Gillam from B.P. It was then the turn of the Premier who pronounced the the project officially open, uncovering an inscribed plaque made from Olympic Dam produced copper to mark the occasion. To complete the official programme the Resident Manager of Olympic Dam Operations, Bob Crew thanked those present for their attendance.

Thirteen years since the discovery and with over seven hundred million dollars spent, production was officially under way. It was providing on site employment for eight hundred people and sustaining even more in the city and beyond. It was only a small operation considering the enormity of the mineralised deposit and was certain to grow when the time is right. There were now over thirty kilometres of underground roadway tunnels in the mine with many more to come. Now in addition to the once remote Roxby Downs station homestead there was the brand new town of Roxby Downs with 480 houses on the maps of the state.

At the end of but not included on the official program was an unexpected and very loud explosion accompanied by an impressive burst of flame from the nearby smelter. Even though it was Guy Fawkes Day it was an unintended event and fortunately no one was hurt. Back on the buses, as we passed the smelter we caught a quick glance at the continuing glow from the mishap. The scene was recorded by an anonymous resident cartoonist on a photo-copied sheet which appeared early the following morning titled 'THE O.D.P. GRAND OPENING and FIREWORKS DISPLAY.' This imaginative drawing also highlighted the friendly rivalry between the, on this occasion, 'holier than thou' mining staff headed by Andy Cullam and the suffering metallurgical staff headed by Ian Lawrence. This seemed to be typical at most mines and the regular banter and chiaking between the two groups indicated high morale. On this occasion Terry Dwyer who was ringmaster for the day's on-site organisation was not left out.

We left the Metallurgical Plant site through the now operating vehicle wheel wash at the gatehouse on the boundary of the 'dirty area.' It was then down the 'Main North Road' (Do they still call it that?) and the sealed Olympic Way the 17 km to the Roxby Downs Township. Luncheon would be served in the large Sports Stadium at the Recreation and Sports centre. Once again the colour coded area as illustrated on our personal itineraries showed

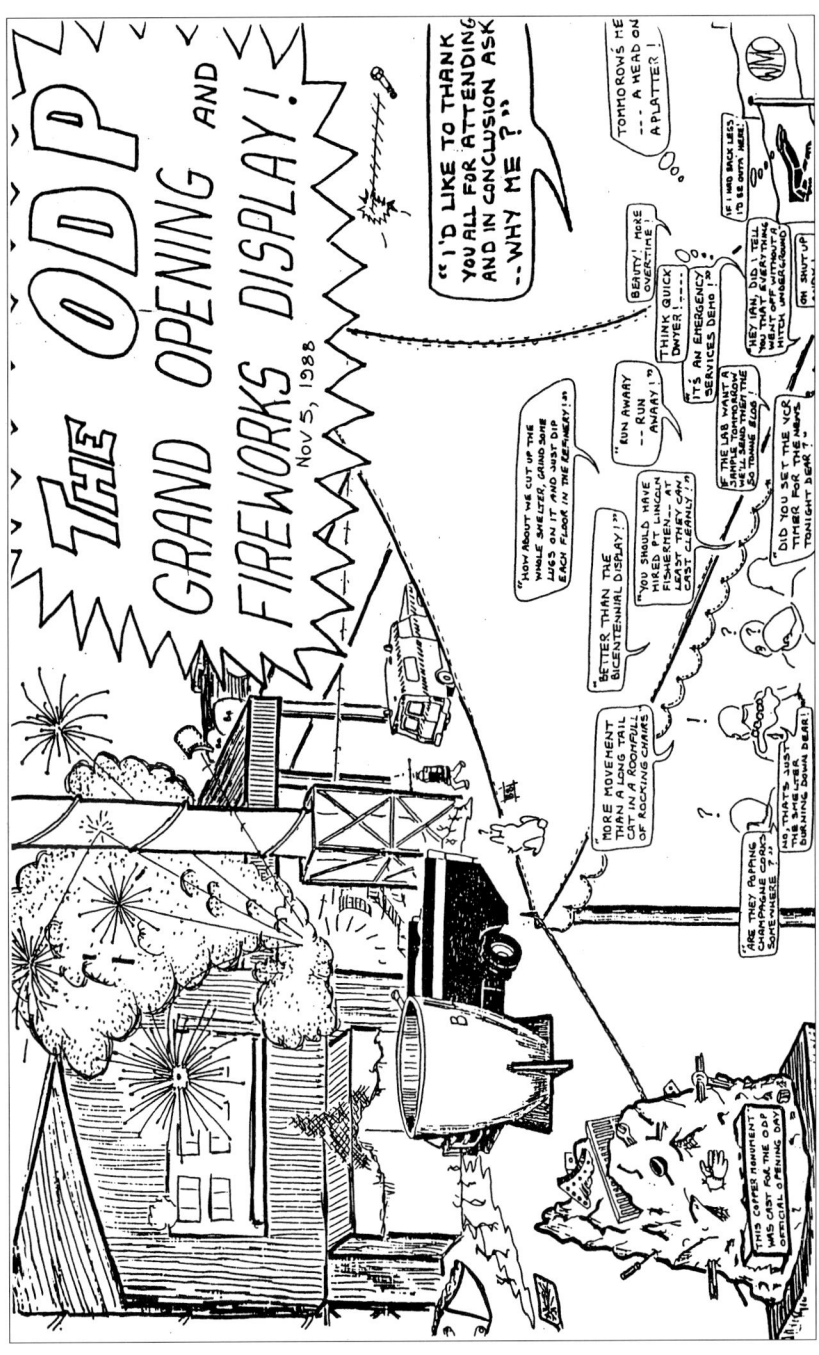

us the tables set aside for those of us on Blue Bus No 9. The meal was very worthy of the occasion and left some free time after to mix socially and explore the town centre. I was particularly glad to be able to catch up with Mick Collins and others who had been the earliest residents at the village.

It was then on to the buses again all too soon for the run back to Woomera and the waiting aircraft. The progress on the new road was now starting to show up having got under way six months earlier. Fortunately the contractor managed to have the roads from the mine to just south of the town completed and sealed for the big day. The rest of the construction and sealing was expected to be finished by mid 1990.

Back in Adelaide in the late afternoon it was a quick change and again out to dinner. This time it was to entertain company visitors from Western Australia who had come over for the opening including my long term friend and boss, Peter Webster, and his boss, and wife, Rita.

At the time of the official opening I was only spending a few days a week on work associated with Olympic Dam project. Peter had required me to go to Perth a number of times and I had also just returned from a week in Florida visiting Phosphate mining and treatment operations. This was to look at their methods of disposal and stacking of gypsum tailings to see whether they could be applied to studies for a WMC project. A few weeks later I was asked if I could help out with the fitting out of new offices of the WMC staff in Melbourne.

From then until July the following year it was commuting to Melbourne for three or four days each week. The remaining time was spent on continuing studies for expansions of production at Olympic Dam and monitoring the new roadworks.

Chapter 31

THE FLOODING RAINS

I was in Melbourne in mid March of 1989 when I received a call from Bill Rymill at Olympic Dam to advise of unprecedented rainfalls. In a period of a three days there had been a total of over 320 millimetres or 13 inches. This was well beyond the limits that had been adopted for drainage design on the project. Long time records in at nearby properties had indicated that the maximum falls ever received in any one month had been less than half of this. The average annual fall over the period of records was in the order of 150mm or 6 inches. It was obviously a very significant event and I was anxious to check out how our works had fared.

The word was that there had been ponding around the shaft area and the stormwater ponds at the metallurgical plant were full to the brim. However they had not lost a minute of production at the plant so we must have got some of our drainage design right. The town was also high and dry but the low lying areas on the golf course were under water. There had been extensive ponding along sections of our original road to Woomera, and deep flooding closing it to traffic to the east of Phillip Ponds. Sections of the new road under construction had been damaged. The airstrip had been surrounded by water but remained in use enabling bread, milk and perishable supplies to be ferried in from Woomera.

The Airstrip after the 1989 Floods. Bluey Lavrick and Pilot Ian Brown at Terminal where water was a metre deep. High Water mark shown by line of debris along edge of runway.

The next day it was back from Melbourne to Adelaide in time to catch the Friday afternoon flight to Olympic Dam. Looking out of the aircraft as we passed Woomera half the country appeared to be covered with sheets of water. On landing we could clearly see that the high water mark of flooding had reached the very edge of the sealed runway. The unsealed shoulders of the strip had gone under but the runway seal had remained high and dry. To the side the apron, terminal and refueling area had been flooded to a depth of about a metre. Bluey was on the spot and pointed out that the water had drained away relatively quickly into the porous borrow pits along the edge of the strip. A channel had also been quickly excavated through the dunes into the adjoining swale which speeded the process. The terminal area had now dried out and was back in limited use other than for the refueling pumps and underground fuel tanks.

All of the storm runoff from the village single quarters, van park and roads had drained towards the borrow pits along the edges of the airstrip as expected. When these filled it backed up flooding to a straight line along the very edge of the runway, and submerging the terminal area. All in all there was very little damage of a serious nature.

Bill Rymill collected me and we headed to the mine where the ponding had risen to within a metre or so of the Shaft Collar. There had been minor flooding on the road into the Core Farm. All of the ponds excavated for catching the storm runoff were full to overflowing but all buildings and the Decline Entrance had remained high and dry. The Shaft was in a relatively large catchment and we had been sensitive to the possibilities of flooding. However the likelihood of rainfalls of the magnitude just experienced had seemed to be very remote even considering the heavy falls at the Borefield a few years previously. There were pipelines in place to pump out the stormwater ponds just in case there were further heavy rains before they dried out. Even so we had escaped very lightly from this very unexpected deluge.

At the the metallurgical Plant the sealed roads with their kerbing and channels had been washed clean and sparkling as they drained the area to the pondages. These had been oversized to obtain extra filling to prepare the plant area. This bonus storage had been fully used and there had been no disruption to production. Out at the Tailings Storage the rains had all been contained well within the perimeter embankments. There was no chance to visit the swales to the north where there was a shallow natural shaft and evidence of old sinkholes in the outcropping limestone. These no doubt would have served as a conduit on this very rare occasion when there would have been substantial recharge to the underlying saline aquifer.

The sealed roads in the mine area, village, and from the the mine to the town were relatively unscathed but all adjoining claypans and hollows were covered with sheets of water. We passed quickly through the town hoping to have time to check out the road to Woomera before dark. The new road had been completed for a short distance to the south of the town and from then on the original road was still in use. As we headed south there were wide sheets of shallow water over the gravel in a number of the hollows between the spaced dunes. There was no apparent damage, the pavement was sound, and the new road 50 metres to the left was high and dry.

Fifteen Kilometres south of the town as we neared the end of the sand dunes we could see the flooding to the east in the expansive hollow of Coorlay Lagoon. We took the short walk to the the top of the high, steep slope that bounded the flooded depression. Previous heavy rainfalls had only resulted in a small area at the lowest parts of the lagoon ponding to shallow depth. This time the entire hollow was covered with mud discoloured floodwater over a length of 6km and backing up along the watercourses leading into it. The water was up to 2km wide and to a claimed depth of up to 8 or 9 metres over most of the area.

At the foot of the steep slope below us a narrow sandy terrace had formed an edge to the normally dry hollow. This flat was home to a magnificent stand of large ancient myall trees and lesser shrubs no doubt nourished from the occasional runoff at the foot of the slope. Only the green tips of the highest myalls were now showing above the murky water. It was only a guess how long it would be before the level would fall to expose them totally and if they could stand the long immersion. The net evaporation rate in this (arid ?) region was about three metres a year and there could be a small seepage loss from the floor of the pond. I resolved to check on these myalls whenever I had the opportunity on future visits. Their failure to survive would indicate that flooding to this extent had not occurred in the lifetime of the oldest trees. Their age could then be determined by a count of the growth rings and perhaps give a feel for the return period of such heavy rainfall.

We continued on towards Woomera across the gibber tableland towards Purple Downs and into the next run of very high dunes. The new road formation with its deep sandy cuttings and fills and crushed rock pavement had also been nearly ready for sealing. It wasn't now as it had been badly scoured along the edges in various places particularly where swift runoff was trapped in a channel between the original road and the new one.

There were no culverts in our lightly constructed old road. The flow from the pipes in some of the gullies under the new road had gushed out scouring deep gutters across the original roadway as a trap for the unwary. The new construction had advanced well beyond the last dunes and was high and dry along the long stretch of open gibber towards Woomera. We found we could travel on the new work without marking the surface so we elected to do this

Coolay Lagoon after the big flood. The tops of the very old Myall Trees that died just showing.

in spite of the keep off signs. The Company were meeting the cost so we felt this entitled us to a faster and smoother ride. This section was relatively undamaged but had banked up the water on to the old road at many places.

It was in the deep valley downstream from Phillip Ponds beyond the end of the new construction that the flooding was the most severe. The water running down from Woomera and beyond had spread out over a wide area. It flowed across the old road to a width and depth that had prevented all traffic trying to make their way through it. Most of this flow originated from a gully on the south side of the road at Phillip Ponds homestead. There it was within a few metres of joining the main stream and continuing on harmlessly to Lake Richardson. Instead of channeling it in this direction the new road designers had insisted on installing large culverts to pass it to the north side under the road. There it flooded the low lying area of the wide valley with the only escape back across the road some distance downstream. This was where we stopped, and turned back towards Roxby Downs, managing a second look at the damage along the way before dark.

Bright and early in the morning it was up to check how the town had withstood the rains and runoff. The concentration we had given to a supposedly 'Fail Safe' drainage design seemed to have paid off. The street system taking the drainage from the higher eastern areas to the low land beyond Olympic Way on the west side of town seemed to have prevented any serious damage. The runoff had ended up in extensive ponding over most of the golf course. Some would be reclaimed for re use on grassed areas and the rest should quickly soak in or evaporate.

Chapter 32

TO THE BOREFIELD AND BEYOND

The first opportunity to check the flooding on the Borefield Road and beyond was as a tourist on the long weekend in June. It was a break in the bush after many weeks spent working in Melbourne. There were three nights camped out in among the eucalypts in the dry creek beds around Lake Eyre and one in the comfort of the Roxby Downs Hotel-Motel. It was the opportunity to venture further along the Oodnadatta Track and old Ghan railway. We were also particularly keen to have a look at the 'Blanche Cup' and 'The Bubbler' mound springs which are natural outlets from the Great Artesian Basin aquifers. Springs and seepages occur in their thousands around the southern and south western margins of the basin. They are scattered along an arc from Cloncurry in Queensland, through outback New South Wales to north of Oodnadatta in South Australia. In the early days they formed the only reliable source of water over much of the inland. As well as helping to sustain the early Aboriginal occupants they enabled the early settlers to survive and determined the route of the overland telegraph and the original Ghan railway. The flow from many of these springs and seepages decreased as numerous bores were drilled throughout the basin by the pastoral industry. A number of the springs surface into shallow ponds ringed with reeds and rushes on top of prominent mounds ranging up to many metres in height. The mounds are thought to have formed over very long periods of time by the deposition of cementing salts and sediment from the water and accumulation of wind blown sand and silt. Many had ceased to flow over the ages but the mounds remain as prominent features over the typical barren saline landscapes.

Turning west on reaching the Oodnadatta Track at Bopeechee we bumped along the corrugated gravel in one of Bluey's hire Land Cruisers. After admiring the mass of water in Lake Eyre South we stopped to check out the deserted stone railway station buildings at Curdimurka. The original bore that provided the water for the narrow gauge steam locomotives on the run to Alice Springs was still functional. The high standing water tank looked rusty and a bit unsteady with a distinct lean. A long section of rail line, restored and maintained by the 'Ghan Railway Preservation Society' stretched away to the west. There it spanned across a high steel multi-span bridge over the wide bed of the 'Margaret Overflow'.

An hours drive further west brought us to the turn off to the springs indicated by an inconspicuous sign and marked by lightly indented wheel tracks. A kilometre or two along there were plenty of choices for the best way over boggy looking, bare flats and around stumpy stoney hillocks.

The 'Blanche Cup' mound rose about six to eight metres above the barren sandy clay pans in the shape of a cropped mini volcano. Clear, shallow, brackish water filled the circular pond at the crest and spilled over a narrow natural spillway and into a gutter down the sloping side. There it spread out on the sandy pans into thin films of moisture going nowhere in particular. Some distance across the flats 'The Bubbler' occupied a similar mound. Both springs had been fenced to deny access to cattle that had caused damage in the past by trampling the edges of the ponds. The surrounds of both mounds were lavishly dotted with artifacts in the form of stone chips left by the early occupants of the land.

I had heard of the display the spring at the bubbler could provide from our environmental monitoring personnel. They visited it monthly as part of the routine measurements of the flows from bores and springs throughout and well beyond our borefields. We were eager to see it in action but there was only a minor stirring of the loose brown sand in a small circle on the floor of the crystal clear pond. It was very disappointing. We waited patiently for quite some time and were eventually rewarded. A rounded bubble of sand in suspension rose in the shallow water in the centre of the pond lifted by the rising artesian flow entering the bed from deep below.. The fluid bubble became mobile, circuiting the pond, adopting various shapes, and rearing every so often to form a fluid dome above the rippled surface. We watched for as long as we could spare. It was hard to leave after such a magnificent display. It was also an opportunity to check out the deep pools in the creeks feeding the lake, which were still holding water from the big rains. At New Years Gift Bore two brolgas entertained with their dancing among the reeds and a search for large fossil mussels was successful. Cattle feeding on the lush growth from the exceptionally good season ignored us. The freezing nights under the stars were brightened with a roaring log fire and it was all over too soon.

Returning along the Borefield Road there was still large shallow pools of water covering the low ground in sections between the north-south dunes. Where these covered the pipeline and road a temporary track had been made to one side on the higher ground along the edge of the dune. The migratory birds were making the most of the abundant water with nests in the clumps of vegetation on small islands in the ponds. The rains had also ensured that the dunes were covered with flowers and fresh growth and being north of the "Dog Fence' there was the odd dingo to be seen.

Passing back through the disputed 'Canegrass' depression the whole of the central area was now an extensive lake nearly reaching the road and pipeline on the deviated alignment. Had we remained on the original pegged line on the other side of the 'swamp' that was now a lake, the road and pipeline would have been flooded to a depth of a metre or so and remained inaccessible for up to a year.

My Last Job – Drilling New Bores – Lake Eyre South.

'Old Ghan' Railway Bridge near Lake Eyre South.

Further along as we approached the mine four large wedge tail eagles were perched regally on a dead kangaroo at the very edge of the road. They did not budge and hardly spared a glance as we passed slowly a few feet away and stopped to admire them. Another was playing games with paddy melons. It swooped down to pick one up, then rose to drop it from a height, either for the fun of it, to show off, or possibly to disturb possible prey.

It was hard to have to commute back to Melbourne later in the week even though it was not to be for much longer.

'Blanche Cup' Mound Spring.

'The Bubbler' Mound Spring.

There was cause for another tourist visit in late September when the the region was blessed with an unbelievable mass flowering of Sturt Desert Peas. The thorough soaking from the big rains had livened seed that must have laid dormant for decades. There was scarcely a swale or open stretch of gibber around the mine and town not covered with the rich scarlet of endless blooms.

It was a once in a lifetime occasion not to be missed, with new growth every where.

There were a number of staff changes in 1989 with Peter Webster resigning from his position as General Manager of the Engineering Group in Perth and Doug moving over there to take his place. Mike took the engineering reins in Adelaide and we continued with the endless studies and preparation of estimates for additions that could lead to expanded production. There were twice monthly visits to Olympic Dam for meetings and for inspections of the new road construction. We were also paying careful attention to the monitoring of the borefield performance with regular reporting to the State.

Early in the year, with the town population settling in, there was a push to develop a Racecourse and an area where horses could be stabled. At Bob Crew's request I checked for possible sites and in due course the enthusiasts developed a site among the mulgas and saltbush a kilometre or two south of the town. Bob also thought it appropriate that a cemetery site be defined and set aside to ensure availability when the need would eventually arise. I recommended a site just out of town, which was surveyed, fenced, provided with an imposing entrance gateway and transferred to the Council for the purpose.

It was becoming obvious that sooner or later there would be a major expansion at the mine and the need to supplement water supplies from Borefield B. The Joint Venturers decided that we should obtain the necessary approvals for the route of the pipeline to that area. This was to avoid long delays when the time came as it surely must. We would have also liked to prove the precise location for the future borefield by drilling and testing a number of exploration bores. However test holes would be very deep and very costly so the Joint Venturers continued to take my word that the additional supplies could be developed in the area. I made sure that they were aware of the long lead time that would apply in developing further supplies from that source.

The pipeline route suggested in the EIS (but not covered by the approvals) extended on from the existing borefield to, and followed the Oodnadatta Track to the east, past Hermit Hill, then ran north east to Borefield B. This was indirect, passed close to sensitive sites, and through long runs of exposed rock. The bores were also likely to be further to the north east than originally thought. It now appeared desirable to head there from the existing pipe route about 40 km short of the Gab 6 pump station.

I had selected a possible route by stereoscopic examination of aerial photographs, a check of the available maps, and several visits to accessible spots along the way. In October 1989 Barry Middleton had arranged a field traverse of the line to check it out on the ground. Armed with maps and aerial photos I led a small convoy of four wheel drives across country from the start point at 80 km along the Borefield Road. We were accompanied by Messrs

Reg and David Warren of Aboriginal descent who represented the lessees of the Finnis Springs property which we would cross. Also with us was David Martin, a consulting Anthropologist. We had previously advised the other affected landholders at the ends of the route of our intent. They were co-operative as the company had a long term policy of providing water tappings at intervals along any pipelines we installed.

It was tougher going than it had been along the existing pipe line as there were wide creek crossings and long steep stretches of broken sand country with dense shrubs and bushes. We crossed the wide rugged beds of Gregory creek, and then Screech Owl Creek, both dry after flooding into Lake Eyre earlier in the year. The beds were dotted with numerous small islands and eucalypts buttressed with tangled flood debris. The maps showed that each of these major watercourses was fed from the south by hundreds of small tributary gullies. The catchments extended back 25 to 30 km from our line in extremely rough terrain to a ridge separating them from the ground sloping south to Lake Torrens. The early fencers had sited a section of the Dog Fence along this ridge to help them cross the worst of the rough going.

Barry was travelling in a brand new company Land Cruiser which handled the rough going beautifully, whereas I was getting roughly dealt with in one of Blue's hire vehicles. He was prevailed upon to let me try it out for the day when I happened to mention a sore back and my advanced age. It speeded our progress as I was leading the way and it cruised up the steep sandy slopes that were stopping the older model. It wasn't the easiest country to traverse with a future road and pipeline so we tried to avoid the worst obstacles. By mid afternoon we had defined a line for about 40 km reaching the Oodnadatta Track and the old Ghan line at Alberrie Creek. We had bypassed the old Finnis Creek homestead at a comfortable distance and no sites of significance were encountered.

Continuing on to the north east we made good progress for about 6 km in open country until stopped by the deep watercourse of Poole Creek. It had deep vertical clay banks and it took some time checking in both directions to find an acceptable crossing. We then traversed up along the valley of a tributary gully rising on to higher ground and crossing the head of Morris Creek.

Another 5 km we topped a low ridge and were overlooking the vast plain extending to the white expanse of Lake Eyre in the far distance. The line ahead took us down a gently sloping open valley which looked just right for bedding a pipeline. After four km or so we were threading our way through a patch of mulga. As we broke out of it we were confronted with the barrier of the 'Dog Fence' blocking our way. Looking along the wheel tracks which followed it's line of thousands of unbroken miles was a gate not fifty metres away. It would be the only way through the fence for ten to fifteen kilometres and we had been heading straight for it without knowing it was there. I immediately claimed it was my superior navigation but didn't fool anyone.

It was by now late afternoon and we were pleased to stop for the night in among a sheltered patch of mulga with only a short distance to cover in the morning. The line we had travelled would have its engineering difficulties when the time came to use it. However it was well clear of the one distant location which was to be avoided according to our Aboriginal guides. A bush meal was cooked on a fire of dry mulga with the conventional billy tea and cigarette. It was swags on the ground among the ever present ants and under the brilliant desert stars for the nights rest.

I learned from our companions of some of the native plants from the area were used as bush remedies and took a sample of one succulent that I later managed to grow in Adelaide. There was now only about fifteen kilometres of treeless open country with braided gravelly stream beds to cross to reach our destination. This was the 'Crows Nest Bore' on Mulloorina Station. This would come into view from afar as it was home to some tall date palms which had been planted close to the free flowing bore. The bore had been pouring out hot artesian water at an uncontrolled rate of well over six million litres a day since it was drilled in the nineteen twenties. The artesian pressure at the surface being about seventy metres. The flow on to the barren surface had resulted in a substantial wetland with reeds and rushes which could tolerate the brackish water. Otherwise it flowed to waste except for the small amount needed to sustain the cattle that grazed the sparse vegetation.

It was the very large free flows from this bore and 'Big Bore' twenty km further to the east that indicated that this was the general location for Borefield B.

Leaving the others to travel and define the line to the bore I returned along our blazed line as a final check and on to Olympic Dam. Barry followed up our selection of the route by arranging an archaeologist to inspect the general alignment in company with David and Reg Warren. An initial botanical inspection was carried out by the State Department of the Environment.

The last two months of 1989 were spent mainly in the office providing input to Mike and his team who were finalising the details of the study to increase mine output to 2.4 million tonnes of ore per year. It was a welcome break as I had flown out of Adelaide over forty times in the year, mostly to Melbourne and Olympic Dam.

There was an even more welcome break in the New Year when I was to work at Perth office for a month in connection with a project in the north west of W.A. There was the relaxing trip across the Nullabor to Perth and back on the Indian Pacific train which was a pleasant change from the regular plane flights.

The time over there was marred by the death of John Oliver, my first boss in the mining industry when joining WMC at Kambalda in 1968. He had left the company some years previously, was about ten years younger than me but had been in poor health for some time. We had been close friends, I had chanced to visit him on the night before he died, and was fortunate to be able to attend the funeral with the large number of company staff. I had never encountered a more astute and capable person in dealing with the complexities and co-ordinating feasibility studies for new mining projects. In mid 1980 he had compiled a very detailed report outlining the likely course and timing to get a mine into production at Olympic Dam. It proved to be an accurate prediction of the works and times that were needed to establish the project.

On return to Adelaide, there was a final inspection of the now completed sealed road from Olympic Dam to Woomera in late Febuary. Driven by Peter Hill,and with Don Orchard also from the Mines Department we left the city in the mid afternoon to arrive at Port Augusta well after dark. We had deviated for an inspection of tailings at the Port Pirie smelter on the way.

In accord with Peter's rigid Navy training it was up and away at 5am next morning driving to and through Woomera to stop for breakfast in the bush half way along to Olympic Dam. (Who said Public Servants don't work long hours?) The long awaited bush breakfast was cooked on an open fire and well worth the wait. We had been joined by Bob Spencer from the Highways Department for the inspection of the road. It was a formality as it had been built competently by Macmahon Contractors for the road authority. The Joint Venturers could now pay to the State the final installment of their contribution towards the cost.

There were five more visits to Olympic Dam in the following months. Each time I would look down at Coolay Lagoon as the aircraft started the descent to note the drop in level as the water slowly evaporated. At some stage the salinity of the remaining water must have risen sufficiently to drop out the fine dirt which had discoloured it since the deluge. Miraculously the whole lake seemed to change between trips from dirt stained to crysal clear and vivid blue.

The full lakes had given the residents of the new township the golden and rare opportunity of desert dwellers to indulge in water sports. Boats appeared in back yards and picnic spots next to water were popular. An aboriginal group objected to the use of the nearby Coorlay Lagoon for recreation, claiming it as a site. The activities were transferred to another large lake just off the road about 5 km south of Purple Downs. The landholder, Tom Allison, gave his blessing to put a gate on the fenceline and drive in to the sandy beaches.

The evidence of ample water in the outback was the presence of the large water birds. Pelicans were known to breed profusely at Lake Eyre on the rare occasions that it filled enough with fresh water to sustain fish. The policeman

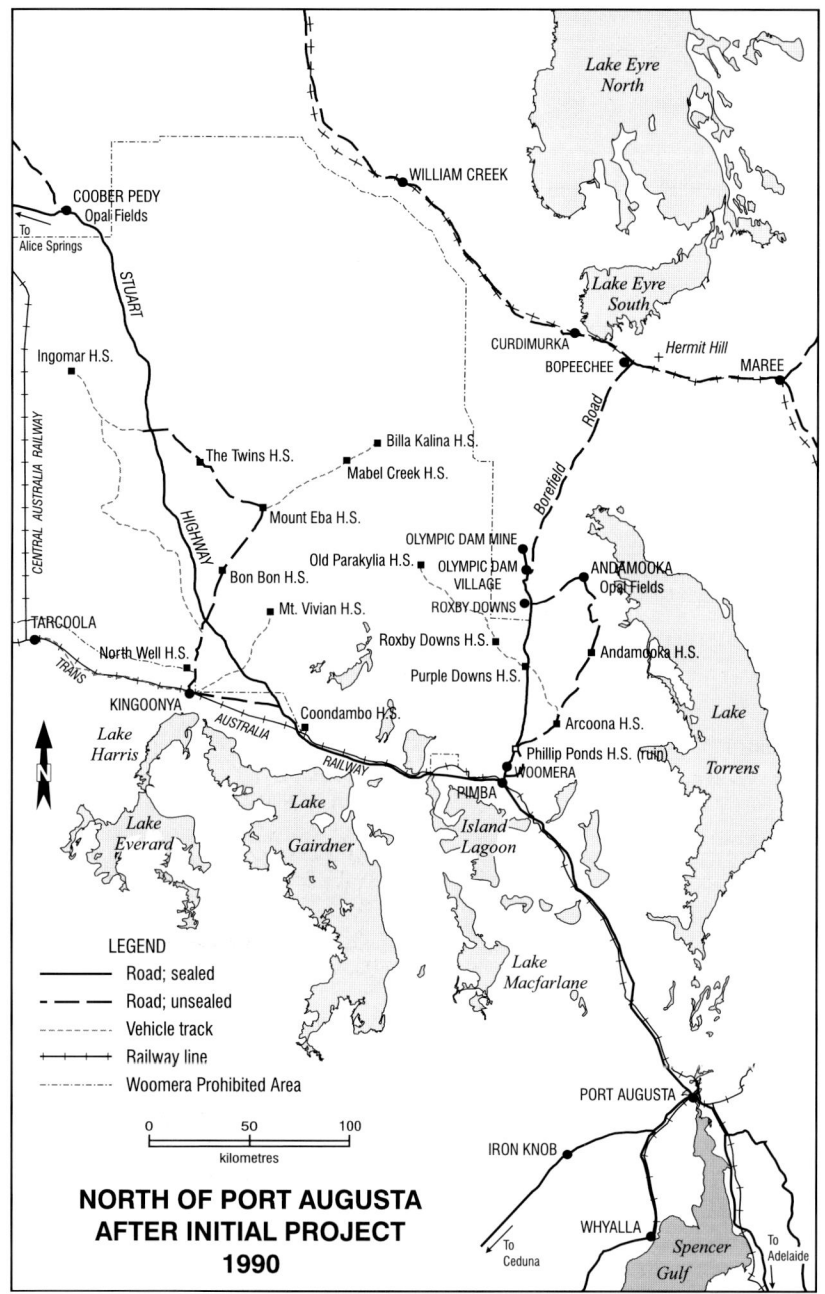

LEGEND

———————— Road; sealed

– – – – – Road; unsealed

··········· Vehicle track

+–+–+–+ Railway line

·–··–··– Woomera Prohibited Area

0   50   100

kilometres

**NORTH OF PORT AUGUSTA
AFTER INITIAL PROJECT
1990**

at Andamooka was amazed one early morning to find that a slightly disabled pelican had landed at the police station, Resourcefully he delivered it to the Olympic Dam Environmental Office where it was taken into care. It was fed with choice fresh fish brought from Adelaide daily and soon became assertive taking full control of their yard and office until evacuated to a coastal area.

In June the Environment Department finally advised that following the inspections of the proposed pipeline route to Borefield B in the previous December that we could now define it with a narrow clearing. In early August we set out to do so with a light dozer followed with a pass with a grader to leave a narrow trafficable track. I set out the line within the approved corridor on engineering grounds closely watched by Barrie Middleton, Frank Badman, his botanist and a consultant archaeologist. When agreement was reached the way ahead was cleared and smoothed.

I was with old friends as Mick and Ron Collins were there to drive the dozer and grader as they had been eleven years ago when they built the original road into Olympic Dam from Andamooka. Peter Watt, a surveyor, followed us along accurately plotting the route and taking levels for future pipeline design. We arrived at the Dog fence on the third day with sixty kilometres of smooth track behind us. I left the others to finish the run to the Crows Nest Bore while I returned to the mine in comfort along the newly blazed line.

Three weeks later after a final visit to Melbourne my employment on W.M.C. staff came to an end. Things were slowing down in the office and in August I was given the opportunity to take retirement a year and a bit early and did so.

There was the usual two month trip away and then a return to Adelaide to enjoy sleeping in and golf midweek without qualms of conscience. It was very pleasant.

It was the official end of 'My Return To Roxby Downs.'

Chapter 33

BACK AGAIN

It was the end of November when the phone rang to see if I might check out some details on the contract for the additional tailings storage I had designed before leaving and which was now to be constructed. So it was back to the airport and a visit to Roxby for two days. As well as tailings there was the detailing of the land in the Village area into allotments, roads reserves and the airstrip, and the 'Annual Water Report' was to be prepared.

These tasks occupied me for a day or two a week without interfering with the golf and it was nice to know I could still be of some help. Then Bob Crew was looking for somewhere to dispose of the massive heap of accumulated mullock (waste rock from underground) that was stockpiled on the surface near the shaft.We agreed that it could be beneficially used as part of the perimeter embankment for the additional tailings storage. As soon as approvals were obtained the heap was on the move giving a head start when the Contract was let to construct the new embankments. 1991 got under way and so did the works on the now approved "Optimisation Expansion" that would increase the mine production up to 2.4 million tonnes of ore per annum. As a part time consultant I was again spending time back in the office mostly in an advisory and supervisory role. Working under Mike's direction and with the past office services still available I started on the infrastructure works required for the expansion.

Firstly it was necessary to find out whether Borefield A could safely supply the additional water required. Cec Forbes was called out of retirement at Port Lincoln. (He had sailed there around the Bight from Perth in an old smallish boat in his first time at sea.) He and John Waterhouse from our long time Groundwater Consultants thoroughly reviewed all the monitoring data and the past drilling and seismic records. The original borefield established in 1987 had six production bores to safely supply the 7 to 8 megalitres a day needed for the initial mine production. The monitoring to the end of 1989 when output had increased to 10 ML/d, showed that with further bores the safe yield from the field could increase up to 15 ML/d. This would be sufficient for the present expansion at the mine and even provide for a possible further stage of development.

Cec and John determined that the best position for new bores should be about 10 to 14 km to the north west of Gab 6 and the pipeline to the mine. One of our monitoring bores, Gab 24, drilled a year or two previously, close to the shoreline of Lake Eyre South, had shown it to be a promising area.

We set about obtaining environmental and other approvals for seismic traverses, and for drilling and testing exploration bores. Assuming success with the drilling, I selected a pipeline route for which we also sought approval. Once again it was agreed that Kinhill would look after the detailed design and contracting for the additions to the existing pumps, pipelines and storages. Tony Reid, Bob Perry and Damien Byrne were soon checking details of the present installations and the future needs.

In addition to this work for Mike, Glen Weir, who had now replaced Bill Rymill as Chief Engineer at Olympic Dam asked for help in urgently expanding the output of the desalination plant.

They needed to add a third train to bring the capacity up to 6 Ml/day. After overcoming some initial problems they were well satisfied with the Reverse Osmosis units that had been installed in 1987. I recommended that the original supplier be asked for a quotation to supply and install the additional plant, to provide all detailed drawings, and to list ongoing prices for replacement membranes and parts. If their proposal was acceptable, then Graham Shelton who was on his staff was more than competent to organize the extension to the building and services.

That wasn't all. Morrie Daw, the Administration Manager was next on the line seeking advice on extensions required to the change room complex at the Metallurgical Plant. This was another job for Kinhill, but again I suggested that Graham be involved and approve the details. In looking after the maintenance of the original change rooms and wet areas he had learned what had been wrong and gave trouble with the existing buildings.

Returning to the water supplies we needed to see how we could deliver at a rate of up to 16 to 18 megalitres a day through the 110 km pipeline from the borefield. It had originally been designed to deliver at a rate of 7 megalitres in an effective 22 hours per day of pumping. The pumps at the start of the line were pushing the flow the full 110 km to the mine. The first twenty kilometres of the line were already subjected close to the maximum recommended working pressure for the ductile iron pipes installed. We needed to more than double the flow rate using the same pipeline as the very high cost of duplication was out of the question. The answer was to install additional pumping units at some point or points along the line to boost the flow along. The Kinhill staff set to with their computers and soon produced the answers. Bigger pumps at the start and a single booster station a third of the way along the line would deliver the flow without bursting the line through excess pressure. It was not going to be an efficient pumping system because of the high friction losses but it was making the best with what was there.

I travelled to site with Tony and Damien to inspect all the water supply installations and to pick a suitable spot for the booster pump station. We also checked out the route of the proposed pipeline to the new bore sites near Lake Eyre South. Dean Arnold who looked after the maintenance and operation of the bores, pumps and pipeline system travelled with us to fill us in on any design changes needed. Then we looked at the installations in the mine area with Bobby Watson who looked after the desalination plant, storages and delivery systems to the mine, plant and town. They were valuable sources of information as to what what was good and what to avoid in the proposed additions.

There was to be a new storage reservoir to be built to store the increased output from the desalination plant. I determined that a capacity of 60 megalitres would be about right to balance out the seasonal fluctuations in usage. It was decided that there should be a plastic liner and also a floating plastic cover. This should eliminate leakage and contamination from soil, birds, dust and airborne debris. Incidently it would also deliver the thousands of years old water from the borefield to the household taps without any encounter with sunlight or the open air. After the new reservoir was built the original potable water pond would be used to store additional process water as a safeguard against interruption to the borefield supply.

We selected a site for the new storage as close as possible to the others and had some test pits dug to check the ground conditions. Although sound we opted for underdrains to go under the plastic liner to detect any leakage that might occur. The dramatic failure when the original process water storage was filling was still very much in mind and a repeat was to be avoided at all costs. The total scope of the works on the water supplies including modifications to the on-site pump stations could now be defined and got under way.

In the Township more single quarters and houses were needed which required a new residential subdivision as almost all of the existing lots had been used. There was now no Hughie so it fell to me to get these under way. Help still being available in the office was invaluable. Louise Bew who knew all the technical terms still found time to do my typing and travel bookings as well as all her other work as secretary to Mike and his engineers. There was also drafting assistance from Vern Egarn who had been there for years and knew my requirements from long experience. He could quickly locate ancient record drawings needed to plan the new works from deep in storage and seemingly lost for ever.

It was mid April before the environmental clearances for the works to extend Borefield A were advised and we were able to have the seismic contractors complete their task. Armed with the print outs from the seismic traverses, maps and aerial photos I visited John Waterhouse at his Perth office. There after a telephone conference with Cec Forbes we selected the sites to drill three exploratory bores for testing. On return to Adelaide and checking the seismic results and proposed drill sites with the Mines Department

Groundwater specialists we finally awarded the contract for the drilling. It was then a visit to the shore of Lake Eyre South, 10 metres below sea level, with a surveyor to peg and record the location for each hole and make any adjustments to the positions as deemed necessary.

The drillers were soon at work and by early June were drilling the second hole. It was a good opportunity to visit and again spend some time touring around and camping out in the bush for an extended weekend. Mike was also visiting the borefield area for the first time so we met him at the drill rig and filled him in on the mysteries of finding water without a divining rod. The exploratory hole at each site was very carefully supervised and monitored by the groundwater consultants. The drill cuttings were collected and lined up in small heaps in sequence of the depth they came from and the depth, thickness and type of rock forming the aquifer was recorded. When drilling was complete the hole was cleared of the heavy mud used to suppress the artesian pressure during drilling. It was then flow tested at different rates and the characteristics of the aquifer and its likely yield determined. The overall information obtained then let the consultants finalize the design for the production bore to be drilled a short distance away. This design would specify the length and mesh size of the tubular stainless steel screen to be lowered into the production bore to the depth of the aquifer. The water flowing in through the screen would then come to the surface through a riser pipe to the borehead with its control valves, pressure gauges and offtake pipework. The exploration drillhole would be retained and used to monitor the affects on the aquifer pressures as water was drawn from the production bore.

The results from the first hole which flowed initially at 7 Ml/day were excellent and the second also promising. The potential safe yields from three new bores would be sufficient to meet the needs for the project expansion. The pressure of the artesian flows would be sufficient to pipe from each bore to a central point without the need for borepumps. This was to be expected as we had the advantage of drilling from a ground level some 10 metres below sea level in the lowest part of Australia.

After some difficulties the three new production bores were completed and tested. Knowing the safe flow rates we were then able to to size the pump station and pipeline back to Gab 6. One complication in obtaining approvals for the pipeline was in the crossing of the old Ghan railway alignment. Although the track was now long abandoned the railways still held title to the right of way and it took time to find someone who could relate to it. Then the Highways Department had to issue a permit to lay the pipe under the Oodnadatta track.

We had added to the drilling contract the provision of a new monitoring bore at Curdimurka. The pipes and valves on the old railway bore were corroded and no longer reliable for measuring any changes in the pressure in the aquifer at this location. We particularly wanted to ensure that the borefield did not adversely affect the pressure in that direction and the mound springs further to the west. A site for the replacement bore was selected a kilometre or so to the east of the old railway station and turned out to be a dry hole. There appeared to be an underground ridge of basement rock in this direction which would extend a barrier between our new production bores, Curdimurka and points west. We drilled again, this time to the north of the station and were successful.

Whenever the opportunity arose during the visits to site I would check out the amount of water in Coolay Lagoon. The continuing evaporation eventually lowered it so that the very large myall trees I had been watching were now high and dry. Unfortunately they were also dead, being unable to survive the long immersion. Barry Middleton was persuaded to have one of the largest chainsawed and the annual growth rings counted. It apparently was a rough count and I subsequently learnt that the particular tree had been about 350 to 400 years old.

This was exciting as it demonstrated that flooding of the magnitude of the 1989 event had not occurred for at least that long and possibly much longer. It was something that merited further investigation and should have been followed up to add to the information on return periods of very high rainfalls in these inland areas.

These high falls resulted in extensive ponding in the innumerable hollows throughout the region. This in turn provided the ideal but very infrequent conditions for recharge of the local groundwater aquifers. Significantly mounding in the groundwater levels in the vicinity of the project tailings storages was detected many months later. Seepage from the storages was suspected as a possibility. However recharge from the north and west where areas of exposed limestone provided ideal condition had to be a likely cause. The initial monitoring of water levels in the early drill holes showed the gradients in the aquifer and the slow natural drainage was from west to east. This was then affected when the shaft sink intersected the aquifer. The subsequent inflow and continuous pumping caused radial flow towards the shaft from increasing distances with a substantial cone of depression.

The increasing use of the company controlled Borefield Road by the public led to an agreement with the State for it to be handed over to their Highways Department. It was now a recognized route to the Lake Eyre region and to the Oodnadatta and Birdsville tracks. To protect the Company interests it was necessary to hold easements over the corridor embracing the road and pipeline. The easements would also need to provide sufficient widths for a future pipeline and power line to the borefields should these be needed in the future.

When Geoff Whitham thought that we had considered all aspects I travelled the length of the road and the borefield with Frank Boulton and the surveyor detailing the specific requirements for the title surveys. This was in September of 1991 and it turned out to be my last official visit to Olympic Dam and Roxby Downs. It was disappointing to see what the Highways Department maintenance grader had already done over a section at the northern end of the road. It had widened and left bare dirt well beyond our neat original six metre width of raised formation bordered on each side with undisturbed vegetation.

There were then only a few other matters in the Adelaide office to occupy part of my time until the end of the year. Included was the next 'Annual Water Report' and a detailed estimate of the works required to drill and prove the availability and location of water supplies from borefield B.

In the New Year my twenty three years with Western Mining came to an end and it was back to the small farm in the mountains of Victoria. Since moving to Adelaide for the three months in 1979 I had returned to Roxby Downs and that part of it around Olympic Dam one hundred and ninety six times. Roxby Downs was now firmly etched on the maps as a thriving new township in the dunes that has since expanded as has the massive mining project that gave it life. So has the village of Olympic Dam with it's Airstrip and supporting facilities for the mine.

The Joint Venture between W.M.C. and B.P. that first brought the Olympic Dam mine into production in 1988 came to an end in 1993. The operation is now wholly owned and managed by W.M.C.

The 'Optimisation Expansion' in 1991 followed by another in 1995 (Twenty years since the discovery drill hole) lifted the mine output to 2.9 million tonnes per year. These included the sinking of a second shaft and the installation of a new mill. The annual mine production was lifted to 84,000 tonnes of copper, 1500 tonnes of uranium oxide, 25,000 ounces of gold and 400,000 ounces of silver. There were some increases in the town population but the installations at Borefield A adequately met the additional water requirements.

A further world scale expansion was completed and officially opened in March 1999 at a cost of $1.9 billion. This time it became necessary to develop Borefield B and install a new pipeline to bring the additional water to the mine. A state of the art underground rail haulage system was fitted in the 150 kilometre maze of tunnels to bring the ore to a third shaft for hoisting to the surface. There the greatly enlarged metallurgical plant is designed to produce 200,000 tonnes of copper each year as well as significant increases in the other products. And it is likely to do so well into the next century and even beyond.

The township of Roxby Downs is now home to about 4000 people with a high birth rate of about 100 babies a year who will regard it as their home town. The Roxby Downs Station homestead is no longer on an island and now not nearly as isolated or lonely as it was when I first went there in 1947.

Appendices

THE GREAT ARTESIAN BASIN

The Olympic Dam Mining Operation and the Township of Roxby Downs are totally dependent on the supply of water drawn from The Great Artesian Basin. (G.A.B.) This vast sedimentary basin of 1.74 million square kilometres covers parts of Queensland, New South Wales, the Northern Territory and South Australia.

Opponents of the mining, principally because of the uranium being produced, have used ill informed arguments critical of the use of this water and advocating conservation of the resource. It is universally accepted that the water stored in the basin and flowing very slowly through the porous sandstone strata is derived from rainwater that has entered where these rocks outcrop at higher levels around the boundaries. Also that the aquifer is similar to a full tank with surplus surface flow periodically available at the inlet beds and that the only water able to enter is that required to replace the amount taken or leaking out.

Prior to European settlement and the sinking of wells and drilling of bores the recharge into the aquifer balanced the losses by way of flows at springs and upward seepages to the near surface sediments. These vertical seepages through the highly impervious and thick rock overlying the porous layers probably accounted for the bulk of the losses. Although upwards seepage would be at an extremely slow rate the immense area of the basin and the high shut in pressures would have resulted in enormous losses in total. This slowly rising seepage as it neared the arid surface in minute amounts evaporated leaving it's contained salts behind in the ground. These either accumulated in layers some metres below the surface or were absorbed into the saline, unconfined, near surface aquifers that overlay many areas of the artesian basin. It is estimated that the annual GAB flow to the 18% of the basin that is in South Australia brings with it over 250,000 tonnes of salt.

Since the 1880's about 4700 flowing bores and 20,000 non flowing wells have been constructed to draw water from the G.A.B. resulting in progressive reductions of pressure in the aquifer. The lower pressures would have lead to a reduction in the quantities of water lost to vertical seepage, and also to an increase of recharge entering at the intake beds. It is certain that a substantial amount of the flow from the bores came from the reduction of the losses from vertical seepage. A downside of the largely uncontrolled flows from the bores was a reduction or cessation of flow in many of the mound springs along the boundary of the basin.

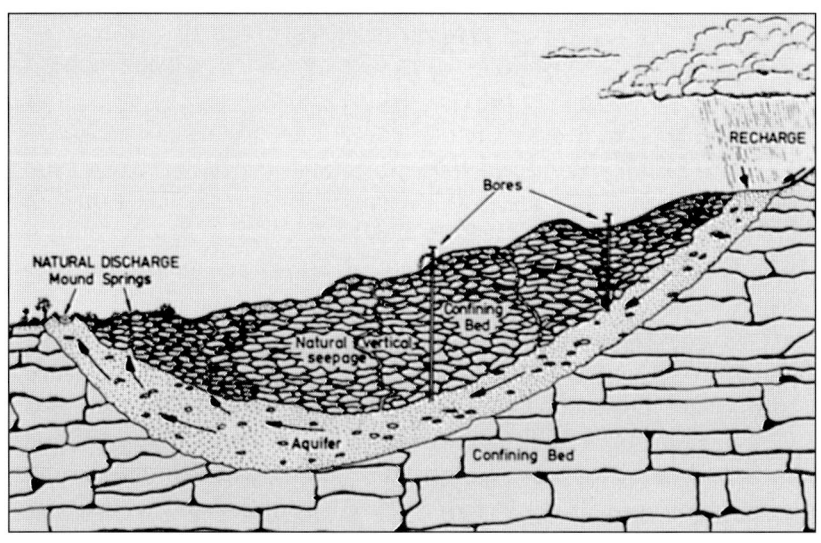

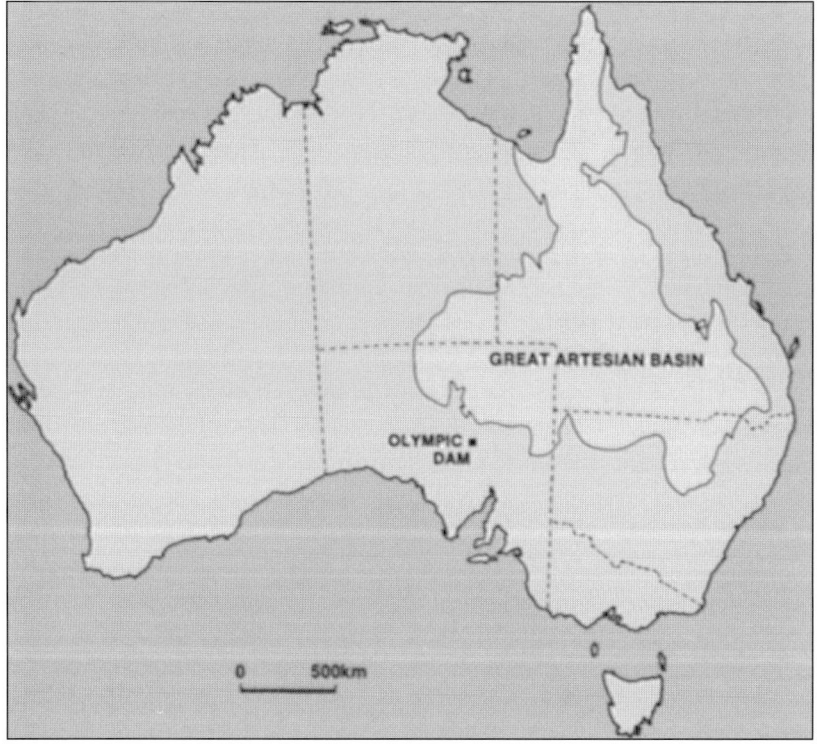

The water obtained from the G.A.B. other than close to the recharge areas contains dissolved salts and has a high residual alkalinity that renders it unsuitable for irrigation. The bores were developed mainly for stock watering with most flowing freely at high rates when only a small amount was needed. The Authorities current concentration on capping and controlling the flow from these will result in a gradual restoration of higher pressures in the aquifers.

However this will have negligable affect on the total amount of water stored in the basin and will very likely result in a reduction in the quantity of recharge.

The total water stored in the G.A.B. is estimated to be hundreds of times more than that held in all of the surface water storages in the country. However there is little or no additional use that can be made of it in it's raw state other than for the limited needs of the mining or other industries in the sparsely settled inland. It requires expensive desalination treatment to produce potable quality supplies for domestic use, and is at great distances from the centres of population. The only settlements and or industries in South Australia that could consider supply by desalination are the special case mines or townships like Olympic Dam and Roxby Downs.It may be that uses for the water in it's brackish state for aquaculture or wetlands for tourist purposes could be developed. The additional flows to springs would also be of conservation value. Otherwise the capping of bores and reductions in draw, although building up pressure and being admirable, will not result in any water being 'saved,' nor any increased beneficial use, nor be of any real benefit to existing users.

5th FIELD SURVEY SECTION - ARMY SURVEY CORPS - 1947.

There were ten men under the command of Major Lockwood, then Major Relf, when the 5th Field Survey Company was assembled in Adelaide in 1947 and commenced mapping for the 'Rocket Range' to be established in the north of South Australia. The members of this small group went their various ways in later years and I have only been able to track the movements of the following:-

LEN BEADELL. Lennie left the Survey Corps to work with the Long Range Weapons Organisation and The Weapons Research Establishment. His exploits included single handedly, first exploring a vast area from Northern South Australia to the West Australian coast. He also managed the on site location and construction of over 4000 km of roads through this unexplored region. He documented his life in the bush very adequately in a number of best selling books and was in demand as a "celebrity guest speaker". He married Anne and they reared a family of two daughters and a son. Lennie passed away on May 12th 1995.

FRANK COHEN. I lost touch with Frank until 1998 when I contacted him at his home in Perth. He stayed with the Survey Corps returning to Woomera, then to Western Australia. He continued using his expertise of survey matters with the Lands Department and in the major mining developments in the Pilbara region. He is a widower living in retirement in Perth.

IVAN MILLER. Ivan left for his discharge shortly after our arrival in the Woomera area and studied civil engineering at university in Melbourne. On qualifying he was employed by a prominent group of consulting engineers in Melbourne and became a senior partner in that organisation.

HAROLD WATTS. Harold returned north with the unit in 1948, and married Doreen Crosby from Kingoonya. They settled in Melbourne where they raised a family, and Harold pursued his career with the Victorian Lands Department. I managed to renew contact with Harold in 1996 but sadly I received a call from Doreen only weeks later to advise me that he had passed away.

OSSIE OSBORNE. On passing through Kingoonya in 1975 and calling in at the hotel I learned that Ossie was managing the North Well sheep station a few miles away. A phone call and Ossie joined us for a meal at the pub, where we discussed old times until late into the night.

BLUEY HUNTER. Blue is reported to have conducted a prosperous private practice in surveying at Darwin.

KEY DATES - OLYMPIC DAM DEVELOPMENT

May 2 1975.	'Andamooka' Exploration Licence granted to Western Mining Corporation Limited.
June 12 1975.	Drill Hole RD1 commenced alongside a stock watering point called 'Olympic Dam' on the Roxby Downs pastoral lease.
July 4 1975.	Drill Hole RD1 intersected mineralisation.
July 30 1975.	Drill Hole RD1 completed. (30 metres of 1% copper.)
Nov 25 1976.	Drill Hole RD10 completed after intersecting the first high grade ore.
Feb 28 1977.	First Study of possible potential of the Olympic Dam discovery commenced.
July 27 1979.	BP Minerals entered Joint Venture with WMC to greatly increase the exploration drilling.
Oct 11 1979.	Sites selected for Camp and Airstrip.
Dec 21 1079.	First Aircraft use Olympic Dam Airstrip.
Jan 2 1980.	Olympic Dam Single Quarters Camp occupied.
June 21 1982.	South Australian Parliament passed the Indenture Agreement.
June 23 1983.	Environmental Approvals for Project received.
August 1983	Anti-nuclear Protesters mass on site.
March 1984.	Pilot Plant operation commenced.
August 1984	Second anti-nuclear demonstrations on site.
Dec 8 1985.	Joint Venturers announce Project Commitment.
March 5 1986.	Construction of initial Project commenced.
Jan 27 1987.	First House at Roxby Downs Township completed.
Febuary 1987.	Water Pipeline from the Great Artesian Basin commissioned.
March 24 1987.	Roxby Downs Community Club Opened.
Sept 1 1987.	Project connected to State Power Grid.
June 22 1988.	First ore milled at Metallurgical Plant.
Nov 5 1988.	Official Opening of Initial Project.
Mid-March1989.	Unprecedented Rainfall totalling about 320 mm over several days cut off road access to the area and filled all lakes and hollows.
Feb 15 1990.	Sealing of road to Woomera completed.
Feb 19 1991.	Joint Venturers announce minor expansion of facilities to increase mine production.
March 1993.	WMC moves to 100% ownership of Project.
April 1993.	WMC announce further works to increase production including a second shaft.

March 1994.   Drill Hole RD982 drilled to 1417 metres.

July 15 1996.   WMC announced plans for major expansion to lift annual production of copper to 200,000 tonnes.

March 26 1999. Official opening of Olympic Dam Operations Expansion.

OLYMPIC DAM - SOME OF THE PERSONALITIES 1979 to 1987.

The following is an incomplete list of the people who lived and worked at Olympic Dam in the years when the political and economic climates made it uncertain if there was ever going to be a mining development. Although I have listed only the males most had wives who also formed a critical and very important part of the workforce.

Mick and Ron Collins. - Earthmoving Contractors.

John Emmerson. - Drilling Supervisor. WMC.

Brett Allen. - Contract Supervisor. Kinhill.

Ron Martin. - Contract Supervisor. Kinhill and WMC.

Jeremy Folwell. - Civil Engineer. WMC Engineering Services. (WES)

Bluey Laverick. - Drill Maintenance Contractor.

Ian (Dasher) Davies. - Electrical Contractor.

George White. - Chief Geologist. Roxby Management Services. (RMS)

Jim Reeve. - Chief Geologist. RMS.

Rob Smith. - Senior Mine Geologist. RMS.

Roly Mortimer. - Supervisor Whenan Shaft Sink. RMS.

Mark Sonter. - Environmental Safety Officer. RMS.

Terry Frith. Whenan Shaft Sink `Blaster' Roberts Construction.

Gordon Bowe - Underground Mine Foreman. RMS.

Gerry Gould. - Senior Mechanical Engineer. RMS.

Bill Rymill. - Chief Engineer. RMS.

Bob Crew. - Resident Manager. RMS.

John Guild. - Senior Drilling Engineer. RMS.

Bill Chandler. - Radiation Safety Officer. RMS.

Tassie Compton. - Camp Foreman. RMS.

John McKinnon. - Civil Services Foreman. RMS.

Ray Spiller. - Core Farm Foreman. RMS.

Alan Jaques. - Underground Mine Foreman. RMS.

Ron Mills. - Underground Mechanical Foreman. RMS.

Bruce Sprott. - Stores Supervisor. RMS.

Mark Busbridge. - Stores Supervisor. RMS.

John Burwell. - Core Farm Foreman. RMS.

Bob Munday. - Drilling Foreman. RMS.
Mick Trasey. - Underground Shift Boss. RMS.
Jack Clarke. - Drilling Foreman. RMS.
Frank Boulton. Civil Construction Supervisor. RMS.
Morrie Daw. Administration Superintendant. RMS.
Alan Phillips. - Mine Project Surveyor.
Barry Hewlett. - Senior Mine Surveyor.
Jim Perkins. - Mine ProductionCo-ordinator.
Ted Charman. - Surface Driller.
Bob Stanfield - Drill Hole Surveyor.
Ian Gilding. - Fencing Contractor.
Tom Burnley. - Plant Operator / Storeman.
Brian Motbey. - First Aid Officer.

THE SHEEP STATION PEOPLE - 1947.

Purple Downs.	Norm Greenfield.
Roxby Downs.	Dave Greenfield.
Billakalina.	Colin Greenfield.
Parakylia.	Walter Greenfield.
East Well.	Joe Stamford.
Mt Vivian.	Fred Stoddart.
Mt Eba.	Bob Crombie.